ALSO BY IAIN McCALMAN

*Darwin's Armada: Four Voyages and the Battle for
the Theory of Evolution*

*The Last Alchemist: Count Cagliostro, Master of Magic
in the Age of Reason*

*An Oxford Companion to the Romantic Age:
British Culture, 1776–1832* (editor)

*Radical Underworld: Prophets, Revolutionaries,
and Pornographers in London, 1795–1840*

The

REEF

The REEF

A Passionate History

IAIN McCALMAN

SCIENTIFIC AMERICAN / FARRAR, STRAUS AND GIROUX

NEW YORK

Scientific American / Farrar, Straus and Giroux
18 West 18th Street, New York 10011

An excerpt from *The Reef* originally appeared, in slightly different form, in *Scientific American*.

Grateful acknowledgment is made for permission to reprint the following material:
Lines from "The Builders," from *Collected Poems* by Judith Wright,
reprinted by permission of HarperCollins Publishers.
Lines from "Australia 1970," from *A Human Pattern: Selected Poems* by Judith Wright
(ETT Imprint, Sydney, 2010), reprinted by permission of ETT Imprint.

Library of Congress Cataloging-in-Publication Data
McCalman, Iain.
    The reef : a passionate history : the Great Barrier Reef from Captain Cook to climate
change / Iain McCalman. — First American edition.
        pages   cm
    Includes bibliographical references and index.
    ISBN 978-0-374-24819-2 (hardback) — ISBN 978-0-374-71170-2 (ebook)
    1. Great Barrier Reef (Qld.)—History.   2. Natural history—Australia—Great Barrier Reef
(Qld.)   I. Title.

DU280.G68 M44 2014
994.3—dc23

                                                                                        2013040660

Designed by Abby Kagan

Scientific American / Farrar, Straus and Giroux books may be purchased for educational,
business, or promotional use. For information on bulk purchases, please contact the
Macmillan Corporate and Premium Sales Department at 1-800-221-7945, extension 5442,
or write to specialmarkets@macmillan.com.

www.fsgbooks.com • books.scientificamerican.com
www.twitter.com/fsgbooks • www.facebook.com/fsgbooks

Scientific American is a trademark of Scientific American, Inc. Used with permission.

1   3   5   7   9   10   8   6   4   2

In memory of
Kim McKenzie, filmmaker,
naturalist, and friend

CONTENTS

Torres Strait

Darnley Is.
/ Mer (Murray Is.)

PAPUA
NEW
GUINEA

Badu Is.
Thursday Is.
Booby Is.
Prince of
Wales Is.
Somerset

Raine Is.

Cape Direction

CORAL SEA

Lockhart
River

Night[s] Is.

Gulf of Carpentaria

Lizard Is.

Great Barrier Reef

Hopevale
Endeavour River
COOKTOWN

Cape Tribulation

Palmer River

Low Isles

Green Is.

CAIRNS

Mission Beach

Dunk Is.

Bedarra Is.

CARDWELL

Hinchinbrook Is.

Orpheus Is.

Palm Is.

Hayman Is.

TOWNSVILLE

Mount Elliot

Whitsunday Islands

BOWEN

MACKAY

Swain reefs

QUEENSLAND

Capricorn-Bunker
Group

Heron Is.

Lady
Elliot Is.

GLADSTONE

Fraser Is.

HERVEY BAY

0   62
Miles

BRISBANE

The
REEF

PROLOGUE

A Country of the Mind

THE AFTERNOON OF AUGUST 25, 2001, is the closest I've come to fulfilling the dreams of my boyhood, when I would lie in bed looking up at the mosquito net and imagine I was Captain Hornblower, sailing a square-rigger to exotic places. And now here I am, sitting on a small beach of scuffed white sand that curves to meet the vast Pacific Ocean. Tamed by the shoulders of the Great Barrier Reef just over the horizon, it kicks up little white breakers that streak toward shore. In the distance bobs a three-masted bark, the HMS *Endeavour*—a replica, admittedly, but real enough for me.

I'm taking part in a reenactment of James Cook's eighteenth-century voyage through the Reef, which is being filmed as a television series called *The Ship* for the BBC and the Discovery Channel. I can see the pinnace and the longboat crawling over the shallow green bay, each boat supervised by one of the dozen professional "officers" who will lead forty-six volunteer sailors. One officer stands swaying slightly in the prow of the longboat, calling out the rhythm to volunteers pulling awkwardly on the heavy oars. She and the pinnace officer are overseeing their attempts to row out to the ship in batches.

I have to wait on shore for several hours because I'm in the last scheduled batch of putative sailors—part of a special group of "expert" advisers

comprising historians, literary scholars, astronomers, botanists, and Indigenous guides. We've been assigned to the mizzenmast, the least lofty
of the three masts. We're generally older and more sedentary than the
other volunteers—all of them lithe and lissome young adventurers from
Britain and the United States—so this will presumably be the least testing of the ship's watches.

I don't mind waiting for the last boat. I sit with my back to a palm
tree, half shaded from the fierce sun, chatting excitedly to a few old
friends. Now and then I take a slurp of tangy milk from a green coconut
that Rico Noble, one of our Aboriginal guides, has given me after kindly
lopping off the top with a machete. I'm mentally reenacting another
favorite boyhood scene, from R. M. Ballantyne's *The Coral Island*, in
which Peterkin, after drinking from a coconut, "stopped, and, drawing a
long breath, exclaimed: 'Nectar! Perfect nectar!' "[1]

True, we're not yet on a coral island, though I can see one shimmer-
ing on the horizon, behind our ship. It is Green Island, complete with
fringing reef and lagoon—the first purely coral island to be recognized
by Cook and his aristocrat associate, Joseph Banks, in 1770. I can't see
any coral from here, but this lovely palm-fringed spot at Mission Bay in
the Aboriginal community of Yarrabah, just outside Cairns, could easily
be the site where our fictional precursors, Jack, Ralph, and Peterkin,
were so providentially marooned.

That was the last moment of unalloyed pleasure I experienced for the
next two weeks.

My first shock was of outraged pride. Scrambling from the longboat
onto the deck, I learned that neither the historians nor the Aboriginal
advisers were to share the privileges of some of the other "experts." We
would work as full-time able seamen—to be freed from sail handling
only when needed to provide a semblance of historical authority for the
television agenda.

Though used to the lowly status of historians within the university
world, I'd not expected such attitudes on what was, after all, a *historical*
reenactment. The captain, Chris Blake, a genuine grizzled seaman with
forty years of square-rigger experience, proved more sympathetic than our
TV masters. He found us a small space at the rear of the ship, normally
reserved for spare sail bags, and he granted us leave, in the odd intervals
between sail handling, to study our own voyage journals and charts, and
to ponder all aspects of discovery and encounter.

Simulating the life of an able seaman on a converted coal bark gave me no time to brood. My annoyance soon turned to terror. Like most tourists, I'd vaguely thought of the Reef as a specific place—perhaps an island resort, a beach, or a section of coral seen while snorkeling. Instead we found ourselves dwarfed by a vast country of sea, reef, and coast.

The Great Barrier Reef is so extensive that no human mind can take it in, the exception perhaps being astronauts who've seen its full length from outer space. Gigantism pervades its statistics. Roughly half the size of Texas, it encloses some 215,000 square miles of coastland, sea, and coral. It extends for about 1,430 miles along Australia's east coast, and encompasses around three thousand individual reefs and a thousand islands. So vast is it, in fact, that it's only since the 1970s, with the establishment of the Great Barrier Reef Marine Park Authority,[2] that a size has been more or less agreed upon. Prior to that, explorers and navigators gave varying figures for its length.

Having to tack our way through such an intricate maze forced us into continual sail changes. I struggled to endure what Cook's veteran salts had taken for granted: working 112 different ropes, hauling myself upside down over the futtock shrouds, balancing over the yardarm to control a thrashing sail while the deck swayed 131 feet below. I ground my teeth on hardtack biscuit that even the reef sharks wouldn't eat, retched on salt pork, and forced down bowel-churning sauerkraut as an antiscorbutic. At night I lay in a hammock a foot wide with a stranger's butt hovering inches from my nose, while the forecastle resonated to the snores of a human bat colony. Along with the sleep deprivation and lack of privacy, my squeamish modern sensibility also had to contend with the shame of public toilets and the petty indignities of naval discipline.

Everything I liked Cook's crew had hated, and vice versa. They'd been haunted by the thought of a coral "labyrinth" and by the terror of drowning, and they fretted about being marooned in a savage wilderness with no signs of cultivation—their signifier of civilization. I, by contrast, longed to jump off the ship and swim in the silky waters around us, to visit the casuarina-fringed cays (small sandy islands) and forested "high islands" sliding past the gunwales, and to bronze my white body in a tropical sun. So irrevocably had the fearful connotations of "wilderness" changed since the eighteenth century that where Banks and Cook saw a cruel and capricious seascape, I saw a paradise. *The Coral Island*, published nearly a century after Cook's Reef voyage, and similar romantic books had instilled in me the idea that beautiful wild places would heal all my discontents.

From my three Aboriginal messmates, however, I began to learn of another, less benign side of the Reef. Rico Noble, an ex-boxer with a shy smile, and Bob Paterson, wiry and serious, both lived in the Yarrabah community from where our voyage had started. Though young they were regarded as elders: custodians of the Gurrgiya Gunggandji and Gurugulu Gunggandji clans respectively. Bruce Gibson, burly, self-confident, and articulate, was head of the Injinoo Land Trust farther north on Cape York Peninsula. He, too, was an elder, of the Guarang Guarang clan, and was keen to develop an ecotourism business for his people.

As a member of the mizzenmast watch, I spent a great deal of time in the company of these three, and they laughingly nicknamed me "the old fella." Overhearing my complaints of tiredness one evening, Bruce advised me to stop cramming myself into a hammock and join them on thin mats rolled out on the timber floor. From then on I slept more comfortably, rolling with the rhythm of the ship. I had the additional pleasure of listening to their soft conversation each night.

Like so many Australian Aborigines, Rico and Bob—in particular—were nostalgic for their original homelands, located elsewhere on the Reef. Being separated from their country in this way was unbearably sad. Their ancestors and families had been forced out of these heartlands— the geographical, cultural, and spiritual places of origin that had once defined their identities. As the great Australian anthropologist W. E. H. Stanner long ago explained: "Particular pieces of territory, each a homeland, formed part of a set of constants without which no affiliation of any person to any other person, no link in the whole network of relationships, no part of the complex structure of social groups any longer had all its co-ordinates." Losing one's country, he said, could induce "a kind of vertigo in living." [3]

Rico's and Bob's clans had lost their jurisdiction over stretches of sea as well as coast: they had always treated beach, sea, and reef as inseparable elements that flowed into and over one another. "Country" denoted for them not just a particular geographical environment known and cared for in every detail, but a cultural space alive with stories, myths, and memories. It furnished food, drink, and shelter, as well as every sort of sustenance for the mind and spirit. Even so, they spoke about these animated places not in tones of hushed reverence, but with an easy intimacy, as if talking about old personal friends.

Most nights I also heard stories of spray-soaked outings in tinnies (small, unpainted metal motor boats used for recreation) to spear stingray, green turtle, and dugong, or to catch barramundi, Spanish mackerel, and trevally. They assumed that these fish and animals were theirs to eat or sell, yet they also expressed a strong connection to them as fellow creatures and a genuine concern for their species' survival. Under existing Queensland and federal legislation in Australia, "limited traditional rights" to marine resources are recognized. In some park zones Aborigines can be issued with permits to hunt dugong and turtle under restricted conditions, though the practice has attracted strong criticism from some environmental quarters.[4]

Critical of Cook's legacy as an imperial invader, each of the three elders had decided to join our voyage to draw attention to their people's struggle to secure land and sea rights. As long as anyone at Yarrabah could remember, Rico told me, the clans living around Mission Bay had used Green Island as a seasonal base for fishing and hunting, yet the community had just lost a legal case claiming long-term association with the cay and its waters because a European farmer once held a lease there during the nineteenth century. Such was Australian law. Now, Rico said, that same law protects Green Island's fancy tourist resort.

On deck, in the slow, early-morning hours of anchor watch, the three men told stories of how their families and clans had been scattered by the frontier expansion that began in the Reef region in the 1850s and which has continued ever since—successive waves of European settlements, institutions, and policies that also wrested children from parents "for their own good." Behind the men's stoicism I glimpsed endless sequences of fracture and migration, of families and friends being shunted between missions, foster homes, stations, townships, prisons, and reserves.

On August 31 the replica *Endeavour* anchored off modern-day Cooktown, where Cook and his crew had come ashore to repair the ship's coral-impaled hull. While our botanists were being filmed foraging for plants, we historians were allowed ashore to meet with local Aboriginal representatives. Bob Paterson introduced us to his famous relative, the MP and Hope Vale elder Eric Deeral, who was accompanied by his daughter Erica.

Eric described how the sight of our *Endeavour* replica in the mouth of the river had overpowered him. He'd felt a direct frisson of empathy with his ancestors across the centuries, picturing them standing on the grassy knoll and watching the strange spectacle of the three-masted bark. He and his clan group, the Gamay Warra, are part of the black cockatoo totem, and a subset of the Guugu Yimithirr people. To support their claim to the surrounding district of Cooktown, Eric had assembled a set of portfolios placing local oral traditions and topographical investigations alongside research done on Western lines, thereby creating an empirically based record of the long-term presence of this tribe and its clans in the area. In 1997 the Guugu Yimithirr of Hope Vale were among the first Aboriginal people to be given legal ownership of their lands under the Native Title Act 1993 that followed the path-breaking Mabo case of 1992, which for the first time gave Australian Indigenous peoples the legal right to own their traditional lands, provided they could prove continuous occupation by their clan or linguisitic group.[5]

Eric and Erica admitted that it was thanks in part to Cook's journals that their claim had succeeded. Eric's understanding of the history of Cook's visit was nuanced and realistic; he did not gloss over the tragedies that many of his people see as its consequence, but he himself no longer felt any anger. After all, he said, grinning broadly, Cook was now helping to repair some of the damage he'd begun.

The Reef presented yet another face to me on September 4 when we anchored off Lizard Island, 150 miles north of Cairns. We'd again prevailed on the BBC organizers to allow us a few hours to visit this crucial site of Cook's original voyage, and after being taken ashore at 6:30 a.m. three of us set off under the guidance of Debbie, a young scientist from the island's marine research station. Debbie invited us to follow her up a steep rocky peak known as Cook's Look.

Apart from a clump of palm trees that had been planted around the resort, Lizard Island managed to resist the stereotyped South Sea images I'd started out with. From a distance, streaked by early-morning mist, it looked bleak and forbidding; close to, it was dry and brown. We clambered over jagged tourmaline outcrops and pushed past gums that had been stunted and twisted by the southeast trade winds and then scorched by bushfires. In between them grew ragged-edged paperbark trees and kapok bushes covered in yellow flowers. Debbie found some tiny green bush passion fruit that we devoured, reveling in the scent and flavor. Clumps of tussock grass brushed at our ankles and two species of doves tried to drown out each other's calls.

That walk proved to be life-changing in two ways: I found the island's land and seascapes achingly beautiful, falling in love with what I now realize is a distinctively northern Reef aesthetic, and I had my first intimation of the threats to the Reef's survival. I'd read a few newspaper stories about stresses to corals around the world, but never taken them too seriously.

Debbie was proud of the efforts of the research station to preserve the pristine character of the local reefs, but had to admit that even with this much care the corals were showing alarming signs of degradation. She doubted their capacity to resist impending forces of destruction that I only later came to understand. What I did gather from the somber tone in her voice was that she and her scientific colleagues at Lizard Island believed the entire Reef system to be under threat of extinction.

When we reached the summit I stared northward to the horizon, where Cook and Banks had first seen the monstrous "ledge" of reefs that threatened to entrap them permanently. The thin, creamy line in the water now looked to me more fragile than fearsome.

We walked in sober silence down the hill to wait for the longboat. I took a quick farewell swim. Gliding over the multicolored bommies—stand-alone towers of coral—I watched tiny pink-and-blue shell fragments pulsing on the sand with the movement of the waves. Goggle-eyed parrot fish flicked out of reach between clumps of emerald seaweed. Suddenly all of this—even the faux Hawaiian resort around the corner—seemed inexpressibly precious.

Since that voyage nearly a dozen years ago I've visited the Reef many times, and as I got to know its seascapes and stories better I fell deeper under its spell. The Great Barrier Reef, as I learned, was built by human minds as well as by coral polyps. To adapt what Robert Macfarlane says in his wonderful book *Mountains of the Mind*, coral reefs are contingencies of geology and biology, "products of human perception . . . imagined into existence down the centuries." Now that we're in the Age of the

Anthropocene, where humans have for the first time begun to influence geological change, this "collaboration of the physical forms of the world with the imagination of humans" has surely never been more important.[6]

This book is a story of encounters between Reef peoples and places, ideas, and environments, over more than two centuries, beginning with James Cook's bewildered voyage through a coral maze and ending with the searing mission of reef scientist John "Charlie" Veron to goad us to act over the impending death of the Reef. It explores how the Reef has been seen variously, and sometimes simultaneously, as a labyrinth of terror, a nurturing heartland, a scientific challenge, and a fragile global wonder. Yet I don't pretend to offer a comprehensive survey of its modern history. Being drawn instinctively to human stories, I've chosen to write a series of biographical narratives—of around twenty extraordinary individuals, men and women, who've shaped our ideas and attitudes to the greatest marine environment this planet has ever seen.

I've focused mainly on three types of people: first, the Western explorers, resource seekers, and scientists who investigated the Reef; second, the Indigenous peoples, and the castaways they adopted, who lived on and managed the Reef's coasts, islands, and seas just before these were overrun by Europeans; third, the romantic beachcombers, artists, photographers, and divers who found creative inspiration in the Reef's beauty.

Some of my protagonists are descendants of people who have inhabited the Reef for at least as long as it has existed in its present form. Others were thrown there by chance, and discovered nurturing and love from the kindness of strangers. Some sought money or power, some fled there to escape civilization's discontents or the guilt of personal crimes. Some were drawn by ambition, revenge, or scientific curiosity, some by the beauty and marvels of the corals, beaches, forests, and creatures. Whatever their motives, they all eventually shared one thing—a passion for this coral country that is like no other in the world.

In the process of writing I've also come to a strong personal conviction. It is only by melding our specialized scientific understandings of the Great Barrier Reef with the ideas it engenders—the sensory, the spiritual, the aesthetic—that we will fully appreciate why it demands we be its global caretakers.

PART ONE

Terror

1

LABYRINTH

Captain Cook's Entrapment

JAMES COOK DID NOT KNOW, on Sunday May 20, 1770, two weeks after leaving Botany Bay on the east coast of New Holland, the western portion of the continent, named by the Dutch captain Abel Tasman in 1644, that the HMS *Endeavour* was sailing into the southwest entrance of a vast lagoon where reef-growing corals began their work. It was a channel that later navigators would call the Great Barrier Reef inner passage. Cook didn't realize that then, and he never would.

The point, obvious enough in his journals, needs stressing because so many historians inadvertently treat this phase of Cook's first voyage of exploration to the Southern Hemisphere as if the Great Barrier Reef we know today already existed somewhere in the back of his mind. As if he unconsciously knew he was about to enter into combat with a constellation of deep-water "barrier reefs" that ran more or less parallel with the Australian coast for some 1,400 miles, creating between them and the mainland a shallow lagoon of uneven depths interspersed with three hundred reef-fringed coral cays and striated with sand, rock, and coral shoals. In reality he sailed unknowingly within the reef lagoon for around 500 miles before he became aware of something resembling a coral "labyrinth." Like explorers before him, he'd had no intimation at all of the possible existence of this freakish phenomenon.

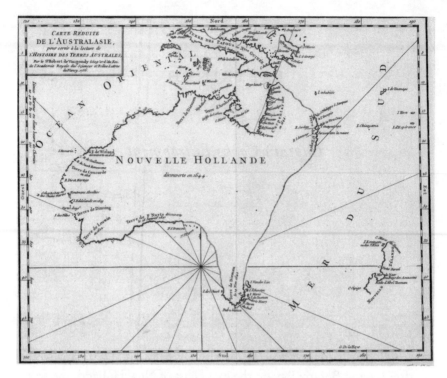

The map used by Cook, showing a still unexplored coastline of New
Holland with Tasmania joined to the mainland. *Carte réduite de
l'Australasie, pour servir à la lecture de l'histoire des terres Australes,
1756* by Robert de Vaugondy (National Library of Australia)

For us to have any glimmer of understanding of the experiences and
reactions of Cook and his crew, we, too, must rid our minds temporarily
of the existence of this vast geophysical phenomenon—a region of land
and sea that in 1770 had never been imagined in its totality by any human being, and that would remain substantially unimagined even after
the *Endeavour* had sailed through it.

Cook had at this point partially completed his mission. He had ful-
filled the orders of the Royal Society to make accurate observations of
the transit of Venus from Otaheite (Tahiti), and was now faced with two
larger and more covert tasks: to best the war-vanquished French by up-
staging their scientific and imperial ambitions in the Pacific; and to discover, chart, and claim for the King of England—with the agreement of
any native peoples—the elusive great southern land that geographers

had so long hypothesized. Having made landfalls on the isles of present-day New Zealand between September 1769 and March 1770, the *Endeavour* had on April 19 sighted land along the coast of what Cook called New South Wales. On April 28 he finally managed to land on this tricky coastline, at what would become known as Botany Bay, a paradise of plants only slightly marred for him by the elusiveness and hostility of the native inhabitants.

Since leaving Botany Bay on May 6, Cook had sighted lines of breakers suggestive of submarine shoals on several occasions, but it was only on the morning of May 20 that he was confronted with a long "shoal" projecting eastward from a finger of land he called Sandy Cape, which forced him to edge northeast for several miles before finding clear water. He named the shoal Breaksea Spit, because after weathering it the ship suddenly entered "smooth water," a consequence of the sheltering effect of the Swain reefs that were far out of sight. Neither was there anything to suggest that the present shoal might be a coral reef rather than an extension of the rocky shoreline, though we now know it to be an extinct coral reef covered in sand.[1]

What Cook actually understood of the origins and character of coral reefs at this point remains uncertain. He'd read the travel account of Samuel Purchas in *Purchas His Pilgrimage* (1613), which described serrated deep-sea coral "ledges," and he'd recently sighted a variety of reef forms in the South Sea Islands, but it was only much later, on his second voyage, that he explicitly echoed the opinion of his onboard naturalists, the Forsters, that coral "rockes" were formed in the sea by "animals." Before this, Cook, like many science-minded men of his time, was probably uncertain whether these protean rocklike objects were plants, animals, or minerals, or a hybrid of all three.[2]

Corals had long been a taxonomist's nightmare, a little-studied phenomenon that early theorists assumed to be some strange sort of plant. In 1724 the Frenchman Jean-André Peyssonnel overturned the work of a colleague in Montpellier with a letter to the Académie des Sciences, arguing for the first time that the so-called coral flower was in reality not a plant, but *"un insect"* that could create bone. His idea was ridiculed until it was taken up some thirty years later by the Englishman John Ellis, who in 1752 told the Royal Society in London that these creatures were "ramified [branchlike] animals," after which his classification became increasingly accepted.

For the deeply practical Yorkshire navigator James Cook, it was more important to know that corals produced vast rocklike edifices that could grow up from unfathomable depths, lurk just under the ocean surface, and sink any ship. At that time it was navigators, more than scientists, who wanted to know what corals were up to.[3]

The Swain reefs responsible for the sudden smoothness of the sea were a collection of massive deepwater coral aggregations some 125 miles to the east that marked the southeastern entrance of the Great Barrier Reef lagoon. In effect, the *Endeavour* had wafted into a vast natural coral basin resembling a woven Aboriginal fish trap; the latter was designed to snare its victims by enticing them into a wide entrance that narrowed suddenly to entangle them, much as the Reef was about to do by veering sharply northwest toward the mainland. As Cook's great editor J. C. Beaglehole observed, anyone telling the captain's story should at this point sound a roll of "premonitory drums."[4]

Cook also failed to sight what might have proved the giveaway presence of the coral cays of the Capricornia group, which were lying over the horizon to the east, some forty-three miles off the mainland. Instead, as they coasted along in a comfortable twelve to twenty fathoms of water with the coast in clear view, they skirted clusters of tall, picturesque islands that Cook named the Northumberland and the Whitsunday groups: these were former mainland volcanic mountain chains that had been transformed into islands by raised water levels and coastal subsidence.

Even the recurring "shoals" surrounding these islands—actually fringing coral reefs—caused Cook no real alarm. Shallows, shoals, and banks held little fear for the veteran sailor who had steered dozens of coal transports like the *Endeavour* through England's treacherous northern coastal waters, and who had navigated flotillas of warships through the rock-filled Saint Lawrence River during the Seven Years' War. Though irritating and, as they increased in incidence, time-consuming, shoals like these could be detected and dodged, provided the leadsman sounded the depths continually and the ship's pinnace was sent ahead to locate deeper channels.[5]

Cook and his young companion, botanist Joseph Banks, did notice that the ship appeared to be entering a distinct new region. The sun was hotter, the air more humid, the sea warmer, the landscape rockier, and the flora more reminiscent of the West Indies. For the first time since leaving Tahiti, they observed palm nut trees and "the true mangrove." These familiar plants convinced Banks that they were departing "the Southern Temperate Zone" and should expect to see more tropical flora. From now on, too, he and Cook would use the tropical West Indies as their template of comparison for the environments encountered. As in the Caribbean, hammer oysters and small pearl oysters were abundant, and both men speculated on the possibility of a future pearling industry for the British Empire. A brief landfall on May 29 further confirmed the similarities with Jamaica, though the lack of water and the presence of barbed grass, clouds of mosquitoes, slimy mangrove mud, and huge tides gave a bleak impression, generating the place name of Thirsty Sound.[6]

The shoal dodging continued as they sailed a slow zigzag course between each new crop of continental islands and the shore. On June 9 they anchored near a small inlet, slightly east of a rocky eminence that Cook named Cape Grafton. It repeated the pattern of high "stony" and "barren" landscapes recently passed at Cape Upstart, Magnetical Island (now Magnetic), Dunk Island, and Cape Sandwich. Here, at the site of today's Yarrabah community, Cook and Banks scrambled up another stony peak to gaze down on yet another mangrove swamp worryingly devoid of fresh water. Spires of "smooks" (smoke trails) indicated the nearby presence of Indigenous people, but none were sighted. That the explorers were being watched, however, is suggested by a faint red painting of a three-masted square-rigger scored on the underside of a barely accessible rock overhang that looks out over present-day Mission Bay.[7] When the *Endeavour* embarked from this bay at midnight on June 10, 1770, under a bright moon and in a slight breeze, Cook had no idea that a chain of coral reefs and cays belonging to what we now know as the outer Barrier lay pincered in toward the northeast, around fifteen miles from the ship. True, he and Banks did note the presence of a cay on a coral reef near their previous anchorage. Cook named it Green Island after the ship's astronomer, Charles Green. Banks suspected that it was "laying upon a large Coral shoal, much resembling the low Islands to the eastward of us but the first of the kind we had met with in this part of the South Sea."[8]

Even so, this isolated coral novelty failed to engender alarm or to change what had become their habitual pattern of sailing off the coast. Night visibility under a glowing moon was good, and a seaman was, as usual, standing at the bows swinging the lead to measure the depth. Cook assumed there was ample time to change course should shoals be indicated. But the retrospective entry in Cook's journal, dated Sunday June 10, serves as our drumroll and presages the end of their innocence, "because," he wrote grimly, "here begun all our troubles."[9]

John Hawkesworth, the clever hack writer who produced the popular Admiralty edition of Cook's papers through which details of this voyage would reach the public for the next eighty years, and who would often insert his own imaginings of Cook's inner state of mind, has the navigator reflect to himself at this moment:

> Hitherto we had safely navigated this dangerous coast, where the sea in all parts conceals shoals that suddenly project from the shore, and rocks that rise abruptly like a pyramid from the bottom, for an extent of two and twenty degrees of latitude, more than one thousand three hundred miles; and therefore hitherto none of the names which distinguish the several parts of the country that we saw, are memorials of distress; but here we became acquainted with misfortune, we therefore called the point which we had just seen farthest to the northward, Cape Tribulation.[10]

A mild scare during dinner when they crossed the tail end of a shoal was quickly succeeded by deep water, so Cook and Banks retired for the night, only to be rudely awakened around 11:00 p.m. when the water shelved suddenly from twenty fathoms to nothing and the ship struck heavily on a reef. Being twelve miles from the shore and still surrounded by deep water, Cook instantly realized that they must have hit coral.[11]

Thanks to Hawkesworth's dramatic account, the crew's subsequent thirteen-hour ordeal, as they fought for the survival of the ship, has become an explorer's classic. We envisage the men, with horror frozen on every face and oaths stifled in their throats, staggering to retain balance as the ship tilts and beats against the rocks with a grating that can be felt through every plank. We watch helplessly while the sheathing and false keel float away in the moonlight; we hear the repeated splashes of more than fifty tons of cannon, ballast, lead, and coal being tossed overboard in a futile effort to float the impaled hull off the coral. Stark disappointment greets the risen tide's failure to reach the ship's bottom, let alone float it free. There remains only the faint hope that the night tide will be fuller.

Hours later there is the sound of the returning tide rushing through the leak, combined with the frantic heaving of successive hands, Banks included, working the three unbroken pumps against the rising water. We feel their exhaustion as they slump on the tilted deck, oblivious of pump water gushing over their bodies. There is a surge of hope on every face as they make an unexpected gain on the leak. Then a last desperate heaving on the capstan and windlass, pulling against the taut anchor chains that radiate from the center and stern, in an effort to jump the ship off the coral. Finally, at 10:20 a.m., the *Endeavour* is heaved into deep water; soon after the young midshipman Jonathan Monkhouse's brilliant fothering (leak-stopping) expedient temporarily plugs the leak. He fills canvas with loose clumps of oakum, wool, and sheep's dung, "or other filth." Cook explains that this canvas must be "hauld from one part of her bottom to a nother until the place is found where it takes effect; while the Sail is under the Ship the Ockham [oakum] &c is washed off and part of it carried along with the water into the leak and in part stops up the hole."[12]

According to Hawkesworth, Cook—tough and phlegmatic seaman though he was—"anticipated the floating of the ship not as an earnest of deliverance, but as an event that would probably precipitate our destruction." Cook assumed, too, that anarchy would ensue as the men sloughed off their naval discipline and fought like beasts for one of the scarce places on the boats, never realizing in their panic that a worse fate awaited them should they actually reach land:

> . . . we knew that if any should be left on board to perish in the waves, they would probably suffer less upon the whole than those who should get on shore, without any lasting or effectual defense against the natives, in a country, where even nets and fire-arms would scarcely furnish them with food; and where, if they should find the means of subsistence, they must be condemned to languish out the remainder of life in a desolate wilderness, without the possession, or even hope, of any domestic comfort, and cut off from all commerce with mankind, except the naked savages who prowled the desert, and who perhaps were some of the most rude and uncivilized upon the earth.[13]

Still, with the leak reduced and hope resurgent, the ship limped for the shore, butted by contrary winds and dodging awkward shallows while waiting for the master in the pinnace to find a suitable channel and a landing place to repair the hull. By Thursday, June 14, he'd discovered a narrow passage leading to a spot on the mangrove banks of what Cook would later call the Endeavour River, the site of modern-day Cooktown. With the wind blowing a gale, the ship "intangled among shoals," and a real danger of being driven onto other reefs to leeward, Cook investigated the master's channel, "which I found very narrow and the harbour much smaller than I had been told but very convenient for our purpose." Even so, they endured three further days of squalls, gales, and groundings on river shallows before they were safely beached.[14]

On June 19, the day after the *Endeavour* had been careened on a rough wooden stage in preparation for repairs, Cook climbed the steepest hill behind the makeshift harbor to get a sense of the countryside where they were marooned. His eyes met "a very indifferent prospect" marked with "barren and stoney hills," salt-infused mangrove swamps, and scrubby trees. He gave no hint, though, of the particular cultural lens that refracted this view: Was it that of a Scottish Enlightenment man of reason hoping to see the cultivated landscapes of civilization, a British imperialist scouting for economic opportunites for future colonists, or simply a nostalgic Yorkshireman yearning for the lush green fields of Great Ayton and the Esk Valley? Perhaps at some level Cook was all of these things, but his journal reveals only an anxious naval professional. For the next six weeks of their land stay, he would fret over the most acute crisis that can face a ship's captain: the survival of his crew in an alien environment, and the feasibility of continuing their voyage home.

To his relief the immediate problem of making the ship seaworthy looked soluble. The carpenters discovered a leak large enough to have sunk a ship with double the number of operative pumps. The *Endeavour* had been saved by a providential lump of coral—the size of a man's fist— that had jammed in the wound to provide a nucleus for the fothered wool and dung. Patching this hole and rebuilding the false keel seemed practicable, but Cook remained worried about the ship's ultimate seaworthiness, especially when it was revealed that the central planking and seams had also been damaged by the strain of careening the ship on land. This resulted in a plethora of new leaks.

On top of all this, the surrounding countryside looked to be devoid of natural resources for the crew. Food was urgently needed because scurvy had struck. Numbers of men were showing the early symptoms of loose teeth, including Banks, who was dosing himself with lime extract. Charles Green and Tupaia, their Polynesian translator, had developed the putrid gums and livid leg spots characteristic of the advanced stages of the disease. Not even the English scurvy expert Dr. James Lind had yet produced a certain cure for this terrible "Explorer's disease." Cook knew from experience, though, that fresh vegetables and fruit always had an ameliorative effect, so he requested Banks to help find these immediately.

The ambitious young naturalist saw this sojourn on land as a chance to gather potentially useful seeds, plants, fruits, and grains for future cultivation, and to discover new specimens to extend Linnaeus's taxonomic scale of the divine order of nature, "in which species had been created and fixed." Banks, who also sympathized keenly with Cook's anxiety about locating nutritious food, alas soon reported that suitable plants were not to be found in this soil, "by nature doomed to everlasting Barrenness." The sailors collected everything they could find, but it proved to be a thin array of cabbage palms, "very bad" beans, fibrous plantains, stone-filled native *wongai* plums tasting like "indifferent Damsons," and a type of wild kale resembling the West Indian cocos. Even Cook, who liked to set an example by wolfing down any form of fresh food, found these cocos roots too acrid to stomach, though he and Banks ineffectually tried to convince the ship's hands that the cooked leaves tasted "little inferior to spinach."[15]

The bizarre animal life around the river, which appeared to be the product of an alternative creation, offered little better. At first the kangaroo seemed a promising source of fresh meat, even if it was an animal that Banks and his colleagues struggled to define. They compared it variously to a greyhound, a giant rabbit, and a local equivalent of the tiny hopping "jerboa" rodent of Africa. Whatever it was, Banks thought it new to science and "different from . . . any animal I have heard or read of." The first specimen killed was "capital eating," but after this the creatures easily evaded Banks's greyhounds in the long grass, and the only other specimens to be shot consisted of a rank-tasting elderly male and a joey carrying little meat.[16]

Ducks, cockatoos, bustards, and parrots were equally shy and elusive: pigeons proved to be the only edible bird slow enough to be shot, but never in sufficient numbers to feed the hands. Neither did anyone manage to kill the yellowish wild dogs or "wolves" that a few seamen sighted. An "opossum" that brought Banks great joy because the French naturalist Georges-Louis Leclerc, the Comte de Buffon, thought them exclusive to America also escaped the pot. One old salt reported seeing an animal "about as large and much like a one gallon cagg, as black as the Devil & had two horns on its head, it went but slowly but I dard not touch it." Hawkesworth pardoned the man's timorousness on the grounds that "the batts here . . . have a frightful appearance, for they are nearly black, and full as large as a partridge." Perhaps the sailors would have eaten this plump-looking flying "devil"—they are a delicacy in Vanuatu today—but nobody managed to shoot one.[17]

After all these efforts, the only person to show a marked improvement in health was the translator Tupaia, who'd set up camp independently on the foreshore and dosed himself on a diet of fish caught in the river. Unfortunately, similar efforts to catch fish for the crew using the ship's net proved too unreliable to make a noticeable difference. This paucity of resources began to look ominous for reasons other than the state of the crew's health. The first was revealed by another hill climb, on June 30, aimed at locating a passage through the maze of reefs encountered on the way into the Endeavour River. "Mr. Banks and I," Cook recorded, "went over to the south side of the River and travel'd six or 8 Miles along shore to the northward, where we assended a high hill from whence we had an extensive view of the Sea Coast to leeward; [this] afforded us a Meloncholy prospect of the dificultys we are [to] incounter, for in what ever direction we turn'd our eys Shoals inum[erable] were to be seen."

Cook and Banks agreed that finding a northward passage "among unknown dangers" seemed the only option, because, as the botanist bluntly put it, the southeast trade wind "blew directly in our teeth." Successive surveys by the master in the pinnace had disclosed a worsening picture. He reported glumly that he could find no clear passage: the ship was blocked in every direction. It seemed increasingly likely that they would remain marooned for many months, until the trade winds altered and permitted them to sail back along the track from which they'd come.[18]

A further disappointment followed their meeting with the local "Indians," who'd been sighted several times but were skittish and evasive about making closer contact. Eventually a series of cautious encounters on July 8 and 9 with small numbers of naked warriors carrying "terrible" lances broke the ice. Both Cook and Banks formed the impression that these tribesmen were lively, intelligent, and athletic, but the explorers' hopes of repeating their South Seas pattern of trading beads, mirrors, cloth, nails, and trinkets for fresh food and vegetables were dashed. Their offerings were received with obvious indifference and then thrown away when the whites departed. Even Tupaia, who won the tribesmen's trust sufficiently to be presented with a few small gifts of food, was no more successful in trading for larger supplies. Only once, when offered a small fish, did the warriors express pleasure and animation, a sign that their desire for nourishing food matched that of the visitors.

The eventual solution to Cook's food crisis came literally out of the blue. During one of the many failed attempts to discover a navigable passage, the master chanced to see large numbers of green turtles basking on the surrounding reefs. After a party of sailors managed to capture a substantial haul of these huge beasts, Banks described the relief and elation felt by all the crew, for "the promise of such plenty of good provisions made our situation appear much less dreadfull; were we obligd to Wait here for another season of the year when the winds might alter we could do it without fear of wanting Provisions: this thought alone put every body in vast spirits."[19]

Should their ship manage to escape from the present coral trap, moreover, captured turtles could be kept alive to sustain the next stages of the voyage. By July 15 Banks was gloating at the turtles' huge bulk— between two and three hundred pounds each—and deliciously flavored fat. "[W]e may now be said to swim in Plenty," he crowed. How they were able to find such a plenitude of turtles so close to the many Aboriginal clans that depended on them is a puzzle. Modern Guugu Yimithirr knowledge custodians suggest that the *Endeavour*'s presence coincided with a period when the clans prohibited the taking of turtle so that their numbers could replenish.[20]

If so, this was an added reason for Guugu Yimithirr displeasure when they saw the *Endeavour*'s decks crawling with enough green turtles to feed their people for a considerable time. On July 19, ten warriors armed with spears boarded the ship in a determined mood. When their leader's request for a gift of one of the thirteen turtles was refused, angry warriors tried to carry two of the massive creatures to a waiting canoe. Several sailors quickly manhandled the men from the gunwales and retrieved the turtles, while Cook tried to appease his visitors with an offer of bread, but "they rejected [it] with scorn as I believe they would have done any thing else excepting turtle."

Infuriated, the clansmen leaped ashore and deftly set fire to the long dry grass adjoining the ship's tents—an act that threatened to destroy the fishing net drying nearby. Cook retaliated by firing a musket loaded with birdshot, wounding one of the offenders. Though the man ran off with relatively little loss of blood, this marked a tragic moment. In a sense, the first British–Aboriginal resource war had broken out, and a fatal pattern of incitement and retaliation had begun.[21]

Yet, if turtles had triggered an environmental war, they also occasioned the first reconciliation between Europeans and Aboriginal Australians. Hawkesworth, with his usual flair for drama, describes at some length the remarkable subsequent encounter between Cook and a group of these warriors:

> I set out, therefore, with Mr. Banks and three or four more, to meet them: when our parties came in sight of each other, they halted; except one old man, who came forward to meet us: at length he stopped, and having uttered some words, which we were very sorry we could not understand, he went back to his companions, and the whole body slowly retreated. We found means however to seize some of their darts, and continued to follow them about a mile: we then sat down upon some rocks, from which we could observe their motions, and they also sat down at about a hundred yards distance. After a short time, the old man again advanced toward us, carrying in his hand a lance without a point: he stopped several times, at different distances, and spoke; we answered by beckoning and making such signs of amity as we could devise; upon which the messenger of peace, as we supposed him to be, turned and spoke aloud to his companions, who then set up their lances against a tree, and advanced toward us in a friendly manner: when they came up, we returned the darts or lances that we had taken from them, and we perceived with great satisfaction that this rendered the reconciliation complete.[22]

This moving encounter was undergirded by a stroke of luck of which Cook was blithely unaware. Guugu Yimithirr oral tradition tells us that the inlet where the *Endeavour* had beached was actually an ancient meeting ground for all the surrounding clans of the district. Here they traded goods, negotiated disputes, dispensed trans-clan justice, and enacted key joint rituals of initiation, marriage, death, and mourning. Failure to share the good fortune of turtle bounty in such a place was thus all the more reprehensible, yet it did not override the clan's ultimate sense of obligation to restore peace.[23]

In spite of the peaceable outcome, Cook later expressed surprise at the sudden deterioration of relations over the issue of turtles. He conceded that the "Indians" might regard turtle flesh as "a dainty"—as did eighteenth-century urban Englishmen—but it never occurred to him that the Guugu Yimithirr might depend on these animals as a staple. Neither did he realize that the "Indians" practiced food sharing and expected it of others, nor that they considered wild creatures like turtles to be the produce of their own estate, which could not be taken without permission by strangers, white or black. Banks at least sensed something of the tragedy that "they seemd to set no value upon any thing we had except our turtle, which of all things we were the least able to spare them."[24]

Thanks to Hawkesworth's literary leanings, Cook is sometimes portrayed as having viewed the Aborigines of Endeavour River through the prism of a "noble-savage" romantic, though it seems doubtful that the practical, taciturn captain would have idealized "primitives" in the manner of Rousseau. Yet Cook was more than capable of recognizing the skill of the Guugu Yimithirr in managing an environment where the far greater part of soil "can admit no cultivation," where fresh water was scarce, and where the seas were filled with terrible hazards such as coral reefs: an environment, in short, that had in a mere five weeks tested the civilized European sailors to their limits. As Cook noted in one of his most quoted passages:

> . . . in reality they are far more happy than we Europeans; being wholly unacquainted not only with the superfluous but the necessary Conveniencies so much sought after in Europe, they are happy in not knowing the use of them. They live in a Tranquility which is not disturb'd by the Inequality of Condition: the Earth and sea of their own accord furnishes them with all things necessary for life, they covet not Magnificent Houses, Houshold-stuff &c, they live in a warm and fine Climate and enjoy a very wholsome Air, so that they have very little need of

Clothing and this they seem to be fully sencible of, for many to whome we gave Cloth &c to, left it carelessly upon the Sea beach and in the woods as a thing they had no manner of use for. In short they seem'd to set no value on any thing we gave them, nor would they ever part with any thing of their own for any one article we could offer them; this in my opinion argues that they think themselves provided with all the necessarys of Life and that they have no superfluities.[25]

Cook praised the Aborigines' ability, with spear and boomerang, to kill birds, fish, and animals so shy that the Europeans "found it difficult to get within reach of them with a fowling-piece." His detailed description of the technology and use of the woomera, or throwing spear, verged on professional awe.

Their offensive weapons are Darts, some are only pointed at one end and others are barb'd, some with wood others with the Stings of Rays and some with Sharks teeth &c, these last are stuck fast on with gum. They throw the Dart with only one hand, in the doing of which they make use of a peice of wood about 3 feet long made thin like the blade of a Cutlass, with a little hook at one end to take hold of the end of the Dart, and at the other is fix'd a thin peice of bone about 3 or 4 Inches long; the use of this is, I beleive to keep the dart steady and to make it quit the hand in a proper direction; by the help of these throwing sticks, as we call them, they will hit a Mark at a distance of 40 or 50 yards, with almost, if not as much certainty as we can do with a Musquet, and much more so than with a ball.[26]

He was impressed, too, by their "facility" in the use and spread of fire, and their skills in building outrigger canoes fourteen feet long, using only shell, coral, stone, and the abrasive leaves of the wild fig tree. He even conceded that their smaller, cruder bark canoes perfectly matched their owners' needs and habitats: "These canoes do not carry above 2 people . . . but, bad as they are do very well for the purpose they apply them, better than if they were larger, for as they draw but little water they go in them upon the Mud banks and pick up shell fish &c without going out of the Canoe."[27]

This last observation was additionally pertinent when Cook and Banks learned from the master on July 19 that he'd found no passage forward or back, northward or southward, suitable for a ship of the *Endeavour*'s size and draft. Banks neatly summarized Cook's dilemma, coining a term to describe the surrounding "reefscape" that everyone, including the captain, would quickly adopt. "We were ready to sail with the first fair wind but where to go?—to windward was impossible, to leward [leeward] was a *Labyrinth of Shoals*, so that how soon we might have the ship to repair again or lose her quite no one could tell (italics added)."

On August 6, after warping out of the harbor and embarking hesitantly behind the pinnace on a northeast course, Cook quickly had to take a fresh survey. He dropped the sails, anchored the ship against the buffeting gale, then climbed the masthead, but still saw no passage. This most decisive of captains was at a loss: ". . . as yet I had not resolved whether I should beat back to the Southward round all the shoals or seek a passage to the Eastward or to the northrd, all of which appear'd to be equally difficult and dangerous." He resembled Theseus trying to circumvent the Minotaur, the monstrous man-bull that lurked in the maze of the Labyrinth, but in Cook's case there was no prospect of an Ariadne to lead them to safety.[28]

Reconnoiters the following day from the headland promontory of Cape Flattery revealed a further shock—what Banks called a "ledge of rocks" or "a Grand Reef" that blocked them from entering the open sea. They'd sighted for the first time what we today call the outer Barrier. On August 11, Cook and Banks rowed to the steepest of a group of three nearby islands in the desperate hope that "the shoals would end." Climbing the highest hill dashed this hope. "When I looked around," recorded Cook, "I discovered a Reef of Rocks, laying about two or three Leagues without the Island, and extending in a line N.W and S.E. farther than I could see on which the sea broke very high." Straining his eyes further, however, he could detect some faint fissures in the long chain of white breakers that might prove to be channels through the Reef. On their way back down to the beach, they named the place Lizard Island, after the giant monitor lizards they saw crashing through the underbrush.[29]

Within hours Cook and his officers agreed they must attempt to navigate one of these small channels into the open sea, rather than risk being "locked in by the great reef," which would likely "prove the Ruin of the Voyage" by forcing them to turn back, lose the prevailing winds to the East Indies, and run out of provisions. On August 13 the *Endeavour* followed the pinnace into a narrow channel earlier reconnoitered by the master. Once through the breakers, they found themselves in "a well growen sea rowling in from the SE," with no ground at 150 fathoms. Once again Hawkesworth imagined the unspoken thoughts that underlay Cook's much terser journal entry.

> Our change of situation was now visible in every countenance, for it was most sensibly felt in every breast: we had been little less than three months entangled among shoals and rocks, that every moment threatened us with destruction; frequently passing our nights at anchor within hearing of the surge that broke over them; sometimes driving toward them even while our anchors were out, and knowing that if by any accident, to which an almost continual tempest exposed us, they should not hold, we must in a few minutes inevitably perish. But now, after having sailed no less than three hundred and sixty leagues, without once having a man out of the chains heaving the lead, even for a minute, which perhaps never happened to any other vessel, we found ourselves in an open sea, with deep water; and enjoyed a flow of spirits which was equally owing to our late dangers and our present security.[30]

Perhaps they celebrated that night with a feast of turtle?

At this point Joseph Banks delighted in the paradox "that the very Ocean which had formerly been looked upon with terror by . . . all of us was now the Assylum we had long wish'd for and at last found." Yet his elation was short-lived, for hardly had the crew finished exulting in the freedom of the open sea than Cook, instead of steering northeast as everyone expected, set a course westward, straight back toward the Reef.

In retrospect this seems an insanely risky act, like a scorched moth returning to circle a flame. Cook later justified himself on the grounds that he was afraid to miss the passage that could confirm whether New Holland and New Guinea were separate continents rather than a single landmass. The chance to make this discovery, which would eclipse the achievements of the mighty Portuguese explorer Fernandes de Queirós, had been on his mind ever since they first sighted New Holland, and he could not bring himself to let it go.[31]

The fruits of this folly were soon upon them. Cook woke at 4:00 a.m.

on August 16 to the sound of the surf "foaming to a vast height." With no wind to give them motion and no ground for the anchor, the ship was carried toward the Reef by the powerful current. Banks recognized this as a unique moment of peril.

> All the dangers we had escaped were little in comparison of being thrown upon this Reef where the Ship must be dashed to peices in a Moment. A Reef such as is here spoke of is scarcely known in Europe, it is a wall of Coral Rock rising all most perpendicular out of the unfathomable Ocean, always overflown at high-water generally 7 or 8 feet and dry in places at low-water; the large waves of the vast Ocean meeting with so sudden a resistance make a most terrible surf breaking mountains high . . . [32]

Two hours later, despite strenuous efforts to tow the ship clear with the longboat and yawl, "we were," Cook observed, "in the very jaws of distruction." Banks was certain their last moment had come: "a speedy death was all we had to hope for."[33]

Just then a few intermittent puffs of wind gave them enough leeway to kick the ship one hundred yards from the breakers, bringing into view a channel through the Reef about a boat-length wide. Cook's immediate attempt to thread this needle was, however, rebuffed by the strong ebb tide, which pushed the ship a quarter of a mile back out to sea. Anxiously they waited for the tide to turn, while the master in the pinnace looked for and eventually located another narrow channel, a quarter of a mile in breadth. Once again hopes rose. "The fear of Death is Bitter: the prospect we now had before us of saving our lives tho at the expence of every thing we had made my heart set much lighter on its throne," wrote Banks. When the flood tide eventually rushed in, "we soon enter'd the opening and was hurried through in a short time by a rappid tide like a Mill race which kept us from driving against either side." The portly *Endeavour* shot through like a nimble canoe. Once they were back within the inner reef lagoon, they dropped anchor in nineteen fathoms on a "Corally and Shelly bottom."[34]

Delighting in the calm, Banks and a few sailors took a small boat to the Reef to hunt for shellfish and turtle. The coral, no longer an emblem of terror, seemed for the first time to be a source of scientific curiosity and aesthetic pleasure. After first collecting three hundred pounds of great cockles for the pot, Banks found himself entranced by "Corals of many species, all alive, among which was the *Tubipora musica*. I have often lamented that we had not time to make proper observations upon this curious tribe of animals but we were so intirely taken up with the more conspicuous links of the chain of creation as fish, Plants, Birds &c &c. that it was impossible."[35]

Though relieved at having orchestrated yet another hair's-breadth escape, Cook's mood was now altogether darker. The inconsistency of his actions in first leaving and then reentering the Labyrinth was obvious to all. "How little do men know what is for their real advantage," Banks reflected, "two days [ago] our utmost wishes were crownd by getting without the reef and today we were made happy by getting within it." This philosophical reflection on the foibles of man appeared to carry no judgment against his captain, but Cook knew he could not presume the same tolerance from his employers in the Admiralty or the gentlemen of the press. The despair and anger that washed over him at this thought led to an unusual spurt of self-vindication.

> . . . such are the vicissitudes attending this kind of service and must always attend an unknown Navigation . . . The world will hardly admit of an excuse for a man leaving a Coast unexplored he has once discover'd, if dangers are his excuse he is than charged with *Timorousness* and want of Perseverance and at once pronounced the unfitest man in the world to be employ'd as a discoverer; if on the other hand he boldly incounters all the dangers and obstacles he meets and is unfortunate enough not to succeed he is than charged with *Temerity* and want of conduct (italics added).[36]

A recent Cook biographer has seen this cri de coeur as a clue to Cook's "deep character" and a revelation of his tendency to self-pity, paranoia, and a "mortal fear of being . . . found wanting," as well as of his overweening hunger for fame. Perhaps this was so, though a historian's judgment is easy to make thousands of miles from the roar of the breakers. To me the moment seems significant more as the disclosure of a profound dilemma: navigating this maze was not only Cook's greatest ever test of maritime skill and physical stamina, but it also confronted him with the explorer's most insoluble moral and psychological nightmare—whether to endanger his men or fail his mission.[37]

From now on he determined to sail northward hugging the coast, "whatever the consequences might be." On Tuesday, August 21, 1770, after a relatively smooth if laborious passage through the remainder of the Labyrinth, including the vortices of currents, shoals, and fringing reefs around the Torres Strait, he was now confident of being "about to quit the eastern coast of New Holland." He therefore landed with a group of sailors and marines on a small stony island to perform a formal ceremony of acquisition: "I now once more hoisted English Coulers and in the Name of His Majesty King George the Third took posession of the whole Eastern Coast from the above Latitude [38] down to this place by the name of *New South Wales*, together with all the Bays, Harbors Rivers and Islands situate upon the said coast (italics added)." That he had not, as his Admiralty orders prescribed, consulted with and gained the prior agreement of the Indigenous peoples must have been an oversight.[38]

There remained some tricky navigation around the barren, guano-covered rock off the tip of Cape York that he named Booby Island, but his crew accomplished it without difficulty. They were by now perfectly drilled in combating the swirling currents and sudden shallows of this capricious sea country. A gentle wind and rolling swell from the southwest convinced the captain on August 23 that the *Endeavour* had passed the northern extremity of New Holland and entered the open sea that lay westward, "which gave me no small satisfaction, not only because the dangers and fatigues of the Voyage was drawing near to an end, but by being able to prove that New Holland and New-Guinea are 2 separate Lands or Islands."[39]

Despite his relief, Cook still felt the need to pen a small apology to posterity. He hoped that a less hazardous passage through the Torres Strait would one day be discovered, never doubting that "among these Islands are as good if not better passages than the one we have come thro'." But James Cook the navigator was exhausted by his battle with the Labyrinth and had "neither time nor inclination" to explore further, "having been already sufficiently harass'd with dangers without going to look for more."[40]

He wanted, in fact, to get out of there as fast as the *Endeavour* could take them, having accomplished his key tasks. Along with assessing and claiming for England the land of New South Wales, which might or

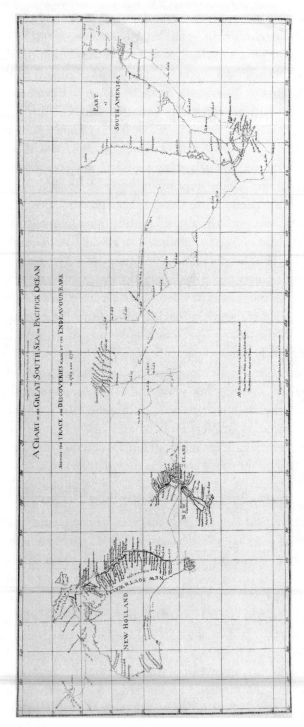

James Cook's chart of the South Sea (National Library of Australia)

might not be a new continent, separate from the westerly land that the Dutch called New Holland, he'd achieved his own personal goal of determining whether or not New Guinea was detached from the northeast coast of New South Wales.

Cook did not know how important it would one day become for British trading ships to have a speedy, thoroughly charted passage through the Torres Strait: for the time being, his protracted route would do. As for the coral Labyrinth, he probably guessed, rightly, that it would interest his masters less as a scientific wonder than as an annoying obstacle for future navigators. But then neither he nor his readers ever realized the true vastness of this coral maze.

These issues were unfinished business, and would one day become the lot of another British explorer-navigator, Matthew Flinders.

2

BARRIER

Matthew Flinders's Dilemma

FLINDERS, WHO WAS NOT YET BORN when Cook turned the *Endeavour* for home, grew up longing to emulate, and then to exceed, his mighty predecessor. By 1802, at the age of twenty-eight, Flinders was commander of the bark HMS *Investigator*, and July 20 of that year found him in Port Jackson, New South Wales, pouring his heart out to his newlywed wife, Ann. He was replying to her twelve-month-old batch of letters from England, which had just reached him. After a grueling survey of a large portion of the southwest coast of New Holland, he was now in the process of refreshing his ship and men in preparation for what he expected to be the most testing leg of a vast journey of scientific discovery: to circumnavigate and survey the great body of southern land known as New South Wales and New Holland, to which he would one day give the name "Australia."

To achieve this he knew he would have to renavigate Cook's Labyrinth, which he now understood a little better than its discoverer, thanks to later charts from two merchant captains, W. D. Campbell and William Swain. Each had, in 1797 and 1798 respectively, revealed the existence of stretches of coral reef far south of where Cook had first encountered them at Endeavour River. Making sense of how these reefs connected with those Cook had seen would be one more challenge in what Flinders

Captain Matthew Flinders, RN, 1814 (National Library of Australia)

privately intended to be a wholesale rectification and extension of his precursor's famed navigation of the east coast of New South Wales.

The young Lincolnshire-born sailor's willingness to bare his soul to Ann came from a profound conviction that the best of modern men combined "the qualities first of the heart, and then of the head." This belief made him heir to two of his century's most influential cultural currents. The heart stood for the cult of "sensibility," a pervasive fashion among middle-class English men and women that required the cultivation of intense and refined personal feelings. A man of sensibility was thought to possess a delicate and elastic nervous system capable of feeling and conveying sympathetic affinities with the emotions and sufferings of all sensate beings. Growing up in the last three decades of the eighteenth century, Flinders and Ann had absorbed the sensitivities of the bestsellers of the day, especially Laurence Sterne's saga *The Life and Opinions of Tristram Shandy, Gentleman* (1759–67) and Ann Radcliffe's gothic romance, *The Mysteries of Udolpho* (1794). Such works had helped shape young Matthew Flinders into a "man of feeling," who, like his boyhood literary hero Robinson Crusoe, combined sensibility with a longing for romantic adventures and explorations.[1]

The head, on the other hand, stood for the rational values and technical attainments of the English Enlightenment that dominated Flinders's chosen profession of naval surveyor. An ambition to excel in navigation and naval discovery had begun with boyhood readings of the great journals of James Cook, whose explorations had added half a hemisphere's worth of new knowledge to geography, navigation, natural history, and ethnography. Trumping his European rivals, Cook had brought new lands and resources into the reach of the British Empire and quickened the march of reason throughout the Western world. In the process, he'd risen from plowman's son to world celebrity.[2]

Gripped by a similar ambition, Matthew Flinders rejected his father's modest occupation of rural surgeon-apothecary to join the British navy at the relatively late age of fifteen. The boy's slight frame belied a fierce will. Advised to concentrate on the skills of navigation and hydrography, Flinders was by 1792 serving as a midshipman on William Bligh's second voyage to the South Seas and the West Indies. Battle experience against the French two years after, followed by a naval posting in the colony of New South Wales, brought promotion and the chance to undertake explorations and surveys for Governor John Hunter.

By the time Flinders returned to England in September 1800, he'd gained enough self-belief to secure agreement from Sir Joseph Banks and the Admiralty for the circumnavigation and survey of New South Wales and New Holland. His orders were to focus in particular on a survey of the Torres Strait and other unknown parts, rather than areas already charted by Cook; to find out whether the east and west coasts belonged to a single continent or were separated by a body of water; and to explore "this, the only remaining considerable part of the globe."[3]

Flinders was anguished to learn from Ann's letters that she had endured a miscarriage, an eye operation, and a nervous collapse, and replying to her required all his considerable resources of sensibility and reason. "Oh my love, my love, how much do I sympathize in thy sufferings," he wrote. So he should, for he was the source of most of her pain. Midway through the previous year, after bewildering Ann with letters that blew hot and cold in line with his career prospects, he'd suddenly rushed her into marriage. Having been promoted to commander, he hoped she'd be able to accompany him to New South Wales, but the Admiralty and Banks vetoed the idea. Rather than risk his new command, he immediately sent his bride of three months back to her parents and departed for New South Wales.

News of Ann's subsequent breakdown was painful enough; worse was her accusation that it had resulted from his "poor proof of . . . affection." In response he begged her to remember that they could never have lived comfortable, independent lives on his English pay. All his actions had been honest and heartfelt: "Heaven knows," he pleaded, "with what sincerity and warmth of affection I have loved thee." Instead, he argued, she ought to stop treating him as the perpetrator of a "crime," accept what fate had dished up, and restore her health with thoughts of their happy future. "See me engaged successfully thus far, in the cause of science and followed by the good wishes and approbation of the world," he enjoined. It was a prospect that he, at least, found supremely consoling.[4]

There was more than a dash of rhetorical calculation in this, and in his successive letters to Ann. Flinders liked to present himself as a suffering romantic, forced by poverty to chase fame and fortune in a remote wilderness that had parted him from his heart's desire. The vast Pacific separating them became a barrier that could be blamed for their pain. "In torture at thy great distance from me," he wrote, "I lay musing upon thee, while sighs of fervent love, compassion for thy suffering health, and admiration of thy excellencies in turn get utterance." Such oscillations between emotional hyperbole and masculine rationality were typical. Often his two creeds of sensibility and science pulled him in opposite directions, making his actions appear conflicted and cold, but they could also sometimes work together, generating a combination of sensitivity and energy.[5]

No leg of his impending voyage, however, would test the compatibility of these two philosophies more than "threading the needle" through Cook's fearsome Labyrinth and the treacherous Torres Strait.

The isolation and mental strain experienced by British survey captains while extending the empire was known to produce a grim toll of suicides, breakdowns, and heart attacks. Flinders understood the loneliness he would face in upholding a necessarily removed and authoritarian role as the voyage's commander, so, like Captain Robert Fitzroy after him, who chose gentlemanly Charles Darwin to keep him company, Flinders selected a sympathetic companion from outside the naval service to provide an outlet for personal feelings. Given that Cook's 1770 voyage had set a precedent for using Indigenous guides, Governor King, Hunter's successor, was not surprised when asked by Flinders in May 1802 if he could enlist two "natives of the country" as supernumeraries. Flinders explained: "I had before experienced much advantage from the presence of a native of Port Jackson, in bringing about a friendly intercourse with the inhabitants of other parts of the coast."[6]

He was referring in particular to the assistance rendered by a young Aboriginal who'd accompanied Flinders on two local explorations in 1798 and 1799. Bungaree, a "worthy and brave fellow," had migrated to Sydney some years earlier from Broken Bay, near the mouth of the Hawkesbury River, and he helped Flinders calm several dangerous skirmishes, such that "[his] good disposition and open and manly conduct . . . attracted my esteem." Flinders normally reserved such warm praise for personal friends like George Bass, the British explorer, but he'd come to regard his "native friend" Bungaree (whose name Flinders spelled "Bongaree") in the same light.[7]

Flinders's need to maintain the distance of command was all the more acute because his sulky younger brother Samuel, a junior officer on the *Investigator*, was inclined at every opportunity to take advantage of their fraternal relationship.

One outlet for the commander's natural warmth was his famous fat black cat, Trim—named after the faithful manservant in Sterne's *Tristram Shandy*. That Trim was also besotted with Bungaree further cemented Flinders's affection for his native companion: "If he [Trim] had occasion to drink, he mewed to Bongaree and leaped up to the water cask; if to eat, he called him down below and went straight to his kid [kit], where there was generally a remnant of black swan. In short, Bongaree was his great resource and his kindness was repaid with caresses."[8]

Bungaree's sensitivity was matched by his courage. On July 30, 1802, a week after departing Port Jackson, the *Investigator* and its supporting brig the *Lady Nelson*, captained by Lieutenant John Murray, arrived at Hervey Bay. This lay just off Great Sandy Isle (today's Fraser Island), at the southernmost entrance of the Great Barrier Reef. The following day, Flinders and Bungaree strolled together along the beach toward Great Sandy Cape, while the remainder of the shore party—botanist Robert Brown, gardener Peter Good, mineralogist John Allen, and illustrators Ferdinand Bauer and William Westall—walked in the opposite direction to collect specimens. The captain and his companion were immediately confronted by a group of spear-carrying Aborigines, waving tree branches and gesturing angrily for the strangers to go back. Bungaree's response set a pattern for the rest of the voyage. He stripped off his clothes, dropped his spear, and walked steadily toward the warriors, chanting words of peace in his own language.

Flinders's susceptible heart melted at what followed: "finding they

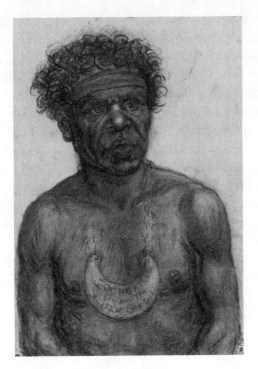

Boongaree by Pavel Mikhaylov (© 2013, State Russian Museum, St. Petersburg)

did not understand his language, the poor fellow, in the simplicity of his heart, addressed them in broken English, hoping to succeed better." Though Bungaree's words might have been opaque, his fearless and open demeanor did its work. The warriors allowed him to approach and accepted his proffered gifts with pleasure. Shrewdly, Flinders followed these overtures by hosting a feast of dugong.[9]

Thanks to Bungaree, Flinders's methods of establishing good relations with Aboriginal warriors always went beyond the formal proprieties suggested in his sailing orders. In stark contrast to Cook's experience at Endeavour River, the goodwill instilled by Bungaree's courage and dexterous displays of spear throwing, accompanied by prolific gifts of hatchets, woolen caps, and mirrors, usually gave rise to a spirit of reciprocation. At adjacent Keppel Bay, for example, the sailors met another band of initially hostile warriors, whom they similarly befriended.

A few days later, on August 15, two sailors strayed from the rest of their party at dusk, and with the sudden plunge of darkness became lost in a morass of muddy shoals and tangled mangroves. The following morning their worried shipmates spied a group of twenty-five Aborigines walking along the beach in the company of two ludicrously muddy figures, "with clothes all rags without shoes and stockings." The warriors had rescued the two sailors, warmed them by a fire, fed them with broiled ducks, and led them at daybreak to the boat. Reflecting later on the character of the region's Aborigines, Flinders was unequivocally positive: "Of their dispositions we had every reason to speak highly."[10]

On their many excursions together, we may guess that Bungaree confided to Flinders something of his own customs, ideas, and practices. Flinders certainly showed an unusually keen understanding of the environmental skills and cultures of Aboriginal people. At Wide Bay, for example, noticing denser than usual Aboriginal populations, he "inferred . . . that the piece of water at the head of Wide Bay was extensive and shallow; for in such places the natives drew much subsistence from the fish which there abound and are more easily caught than in deep water." Such population density also suggested the presence of ample fresh water. He further speculated that the fleshier, stronger appearance of these northern peoples was "a result of being able to obtain a better supply of food with the scoop nets, which are not known on the southern parts of the coast."[11]

When the European sailors' own seine nets proved inadequate, Bungaree showed them other highly effective ways of obtaining fish. His skill at spearing and his unselfishness in sharing his catch filled Flinders with admiration: "it was with much difficulty that his modesty and forbearance could be overcome, for these qualities, so seldom expressed in a savage, formed leading features in my humble friend."[12]

For most of the early part of their voyage, however, it was Flinders's towering ambition to eclipse Cook's achievements as an explorer and surveyor that predominated. Strictly speaking, the *Investigator* and *Lady Nelson* shouldn't have entered the southern end of the Reef at all, let alone attempted to resurvey Cook's original route. Flinders deliberately defied his orders from the Admiralty to focus on the charting of the Torres Strait and Gulf of Carpentaria, because he believed his superior cartographic skills would show up Cook's. In fact, he wanted to survey the Australian coastline so comprehensively that it would never need doing again. Along the way, he intended to find a safer passage through Cook's "labyrinth," and to discover new inland river routes, new animals and minerals, and new sites for settlement. All this had to be done quickly enough to also enable the charting of the Torres Strait and Gulf of Carpentaria before the onset of the northwest monsoon in early November.[13]

The first weeks of the passage seemed to justify his hubris. Cook's rough running survey had been undertaken at speed and without the use of Harrison chronometers, which now enabled the estimation of longitude with far greater accuracy. Instead of taking lunar distances, one need only compare the local time at noon to that of Greenwich and then apply standard mathematical formulas. New sophisticated theodolites also enabled angular distances to be measured along both vertical and horizontal planes. Flinders thus saw himself as one of a new breed of professional navigators who'd turned the art of surveying into a science.

While in bays and harbors, he established a baseline with a theodolite and used the shallow-drafted *Lady Nelson* or the whaleboat to crisscross the waters in a triangular pattern, taking angular measurements of the whole area. When forced to use running surveys he would sail as close as possible to the shoreline—and always closer than Cook—plotting the ship's track between fixed points along an estimated course. Wherever possible, too, he would take extra compass bearings of notable shore features. Finally, Flinders believed his specialized knowledge of the burgeoning science of magnetism would enable him to reduce magnetic variations in compass readings caused by the presence of iron.[14]

As he surveyed the bays and harbors north of Fraser Island during July and August of 1802, Flinders confirmed a succession of corrections to Cook's observations and charts made by subsequent explorers. He also made some satisfying new discoveries. Cook's estimate of the size of Hervey Bay proved to be out by some sixteen miles. At Bustard Bay there were "considerable disparities in their two positions, and the form of the bay did not correspond with Cook's chart." On August 15 Flinders discovered a new port (later named Port Curtis, present-day Gladstone) that Cook had missed while sailing at night.[15]

A little farther on, at Cape Keppel, he found a prospective new settlement site, complete with fertile-looking grass for cattle, "pleasant looking vallies, at the bottom of which are ponds, of fresh water frequented by flocks of ducks," plenty of wood, and excellent fishing. Then, on August 21, he approached what he thought was a deep bight, only to discover another substantial port of "romantic appearance" that had also "escaped the observation of captain Cook." He later named it Port Bowen, and noted that it seemed to have the most abundant fish stock that they'd yet seen. Here, too, the ship's painter William Westall at last found a landscape of towering cliffs and woods sufficiently picturesque to please his Eurocentric eye. And when on September 7 they reached Thirsty Sound, one of the few landfalls Cook had made within the Reef lagoon, Flinders was pleased to report a steadily widening eastward gap between his and Cook's longitude readings.[16]

Yet these achievements came at a heavy cost. By the time the expedition reached Shoalwater Bay in late August, after some ninety miles of arduous coastal and harbor surveying, Flinders had an inkling of the weariness that had pervaded Cook's men. His own officers and crew were exhausted from the endless sail changes and triangulations in choking heat. The ambitious captain began to chafe at the many barriers slowing his progress. His discovery of inland river routes was hampered by endless mudflats and impenetrable mangroves. Outbreaks of fever and diarrhea strained the sailors' tempers and led to a drink-fueled melee in which one sailor struck an officer and had to be flogged. Flinders was himself so weak that he couldn't even match Cook's feat of climbing a nearby hill to take bearings. A new hazard emerged, too, in the torrential currents that swamped several of their boats and carried away the cutter. When leaving Shoalwater Bay on August 27, Flinders wrote in his journal a warning to future navigators to avoid the entranceway he'd named "Strong-tide Passage": currents ripped through it at more than five miles an hour.[17]

At Broad Sound they were stuck for a further twenty days under similar conditions while charting a morass of shoals. Here both Flinders and Murray also had to contend with stupendous ebb tides that varied thirty-two feet or more, often forcing their ships' hulls, when deep channels disappeared, to corkscrew helplessly on the mud shallows. The *Investigator* narrowly escaped capsizing on one quicksand bank, and the *Lady Nelson* grounded on a stretch of muddy gray seabed that cracked its movable keels, front and rear.

Such morale-sapping conditions were exacerbated by episodes of devastating slackness. Twice Flinders returned from inland excursions to find that his brother Samuel had forgotten to wind the chronometers, nullifying ratings that needed to be taken on seven consecutive days. This dereliction of duty threatened "to cripple the accuracy of all our longitudes." Unable to delay longer, Flinders combined what ratings they had and set a course past the distant Northumberland Islands, with the intention of hunting for a passage through the outer reef to the open sea.[18]

Today the wooded islands in this chain are a cruiser's delight, but in September and early October 1802 Flinders knew them only as the start of Cook's notorious labyrinth. Having charts made by Cook, Campbell, and Swain didn't actually speed progress. Flinders was still forced to adopt Cook's painstaking sailing regime of anchoring at night, taking continual soundings, and positioning lookouts on the forecastle and topmast. Several attempts to thread the *Investigator* through narrow channels in the coral saw the ship driven back by tides, whirlpools, and currents that swirled in the funnel-like entrances. A frustrated Flinders found himself echoing his predecessor's complaints: "for by this time I was weary of them [reefs], not only from the danger to which the vessels were thereby exposed, but from fear of the contrary monsoon setting in upon the North Coast, before we should get into the Gulph of Carpentaria."[19]

By October 9 the vessels were so hemmed in by coral that Flinders had to resort to a further Cook expedient. Anchoring until low tide, he set out in a whaleboat to explore the line of exposed reefs in the hope of finding navigable channels. Clambering for the first time onto one of these reefs, he couldn't refrain, even in his present predicament, from expressing a surge of romantic delight at the sight that met his eyes:

> . . . the water being very clear round the edges, a new creation, as it was to us, but imitative of the old, was there presented to our view. We had wheat sheaves, mushrooms, stags horns, cabbage leaves, and a variety of other forms, glowing under water with vivid tints of every shade betwixt green, purple, brown, and white; equalling in beauty and excelling in grandeur the most favourite parterre of the curious florist. These were different species of coral and fungus, growing, as it were, out of the solid rock, and each had its peculiar form and shade of colouring; but whilst contemplating the richness of the scene, we could not long forget with what destruction it was pregnant.[20]

As well as avoiding these beautiful hazards, the two captains had to prevent their ships from being sucked at night into tidal vortices created by the narrow reef channels. The resultant tussles with the ebb tides produced a succession of snapped cables and lost anchors. When the *Lady Nelson* severed both its kedge and bower, Flinders was reduced to sending Lieutenant Murray two grapnels to bind together into a makeshift anchor. These alarming losses finally convinced the stubborn commander that he was wrong "to persevere amongst these intricate passages beyond what prudence could approve; for had the wind come to blow strong, no anchors, in such deep water and upon loose sand, could have held the ship . . . I therefore formed the determination, in our future search for a passage out, to avoid all narrow channels, and run along, within the inside of the larger reefs, until a good and safe opening should present itself."[21]

Two nights later, in illustration of the wisdom of Flinders's new resolve, the *Investigator*'s anchor lifted during the middle watch, allowing the ship to drift unnoticed for two miles before the anchor recaught on the verge of a line of reefs. His fury at how easily this "might have been attended with fatal consequences" redoubled on discovering that Samuel was again responsible.[22]

Still, Samuel was not the only Flinders family member to flout orders. On October 17, after threading through a maze of inner and outer reefs near the Cumberland Islands, the two ships at last experienced an afternoon of clear sailing in deep water. Believing they'd finally found a passage through the outer reef, Flinders abruptly ordered the *Lady Nelson* to return to Sydney. His accompanying letters to Governor King explained that the converted sloop was a poor sailer, ponderously slow and liable to drift to leeward because of its flat bottom and paltry movable keels. Lacking adequate anchors, it threatened to become more a hindrance than a help. In actuality, because Flinders now had little time to reach the Torres Strait before the monsoon, he was ruthlessly removing another potential barrier to his success. The *Investigator* couldn't afford to be slowed by Murray's sluggish little brig.

Sure enough, having rid themselves of the encumbrance, they weathered a few more jagged banks of reef to find themselves again in clear water. By October 21, 1802, the *Investigator* was making brisk time toward the Torres Strait with a comforting sixty-six fathoms beneath its hull. After fourteen days and more than five hundred miles of painful searching, they had discovered a new way through the Reef—later to be named Flinders Passage—which its discoverer estimated to vary between twelve and sixty miles in width.

Despite the need for haste the *Investigator* anchored each night, because of the continued danger of "straggling reefs" in an expanse of water that Flinders called "the Corallian Sea." Nine days of coral-free sailing followed, before they were forced to divert around a large cluster of reefs that lay east of those sighted by two previous British mariners: Captain William Bligh on the *Providence* in 1792, and Captain Edward Edwards on the *Pandora* in 1791. Naming these new coral banks the Eastern Fields, Flinders on October 28 entered an opening in the reefs discovered by Edwards. But heavy banks of monsoonal cloud overhead suggested there would be no time to carry out orders for "a careful and accurate survey" of the Torres Strait.

After another bout of reef dodging, the *Investigator* anchored off the largest of the islands in the Murray group the following day. The crew warily eyed around fifty "Indians" who dashed out to greet them in a fleet of canoes. For most of the previous week Flinders had instructed the hands in cannon and arms drills, but the warriors' eagerness to obtain iron goods in return for bananas, coconuts, plantains, shells, and ornaments suggested that commerce was supplanting head-hunting. Though his vigilance never slackened, Flinders admitted to being impressed by these muscular, ornamented Islanders with intelligent countenances, palisaded houses, cultivated gardens, and swift seagoing canoes.[23]

Departing a day later, they steered as far to the south toward Cape York as the many reef clusters would allow, before eventually coming to anchor near a small island. The initially puzzling sight of scores of giant, one-hundred-pound cockleshells, each positioned under a pandanus tree, proved to be an ingenious method used by visiting Islanders to overcome the lack of water. Flinders noted that "long slips of bark are tied round the smooth stems of the pandanus, and the loose ends are led into the shells . . . underneath. By these slips, the rain which runs down the branches and stem of the tree, is conducted into the shells, and fills them at every considerable shower; and as each shell will contain two or three pints, forty or fifty thus placed under different trees will supply a good number of men."[24]

Naming the cay Halfway Island, they sailed in a southwesterly direction before anchoring on October 31 beside a cluster of islands in sight of what Flinders presumed to be Cape York. Steering carefully among the many fringing reefs for a further few days, he eventually cut through, between Bligh's Wednesday Island and a northwest reef, and entered the Indian Ocean on November 3, 1802. At dusk the same day the *Investigator* reached the Gulf of Carpentaria.[25]

Matthew Flinders had found and charted the present-day Prince of Wales passage through the Torres Strait, which is still favored by most shipping traffic. Conscious of his Admiralty orders to focus on a close survey of the strait, he wrote to explain that "apprehension" of the impending monsoon had prevented this, but he still felt reason to be pleased. He'd beaten the weather and sailed through this coral minefield in only six days, which, allowing for stops, meant that he'd "demonstrated that this most direct passage, from the southern Pacific, or Great Ocean to the Indian Seas, may be accomplished in three days." Sailing between the two great oceans by the usual route around the north of New Guinea took most ships at least five to six weeks.[26]

Yet even while congratulating himself, Flinders was conscious of a new problem. The *Investigator*'s first encounter with the brisk winds and choppy swell of the open Pacific had produced leaking of around five inches per hour, but by the middle of the Torres Strait this had worsened to an alarming fourteen inches per hour. Hoping that the problem was localized, Flinders found a suitable site in the gulf to careen the hull and investigate the issue. The carpenters' reports of November 17 were devastating: the hull was half rotten and worsening fast—planks, bends, timbers, and treenails on both the starboard and port sides were affected. The prognosis was stark: if caught in a strong gale the *Investigator* would founder; if grounded on a shoal it would fall to pieces. Only if cosseted in fine weather could it conceivably last a further eight to twelve months.[27]

The bitter implications of this news still reverberate in Flinders's account many years later: "I had ever endeavoured to follow the land so closely, that the washing of the surf upon it should be visible, and no opening, nor any thing of interest escape notice . . . and with the blessing of God, nothing of importance should have been left for future discoverers but with a ship incapable of encountering bad weather—which could not be repaired if sustaining injury . . . I knew not how to accomplish the task."[28]

With the northwest monsoon now blowing, any hope of sailing down the western shores of New Holland had to be shelved until the wind shifted back to the south. Thanks to Flinders's hubristic resurvey of the area covered by Cook's charts, they were now trapped inside the Gulf of Carpentaria, with no choice but to try to caulk the ship's rotten seams and continue to inch along the broiling northern coasts until the monsoon season passed.

Meanwhile the sailors' health was disintegrating faster than the ship's planks. By the time they reached Arnhem Bay on March 5, 1803, scurvy, fever, and diarrhea were so rife that Flinders, himself crippled with scorbutic ulcers, had to make the agonizing decision to head the *Investigator* to the nearest East Indies port. As they turned toward Kupang in Timor, he asked his surgeon, Hugh Bell, for an overall report of the men's health.

The written reply was blunt to the point of abrasiveness. It had been nineteen months since the crew sailed from England, during which time they'd been given no more than two or three fleeting opportunities to refresh themselves with "animal food," fruit, and vegetables. Over the past eight months, moreover, they had "been exposed to almost incessant fatigue in an oppressively hot climate, as also to an exceedingly deleterious atmosphere since Dec. 16 when the weather became dark and cloudy with thunder, lightning, and rain. The ill effects of this alteration in the weather were perceptible in a short time among the ship's company; a violent diarrhea being produced, attended with symptoms of fever . . ."

On top of this, there were no fewer than twenty-two men with scurvy symptoms, such as ulcerations, loose teeth, spongy gums, livid sores, foul breath, and lassitude. Five of these were already seriously ill.[29]

Simultaneously troubled and angered by the surgeon's tone, Flinders secretly still hoped to resume the survey by procuring a new ship at Kupang. When this failed to materialize, he decided to fulfill his original orders to scout around the region for new harbors or sea lanes that could prove useful to the East India Company. Bell was appalled. Flinders's dilemma, similar to that faced by Cook, of whether to continue surveying or sail straight back to Port Jackson had, in effect, brought his two philosophies into insoluble conflict. The man of feeling fretted about his ailing sailors; the scientific cartographer wanted to stick to his mission. Either way, he lacerated himself for failing to live up to the towering examples of his predecessors:

It may well be, that to leave such a coast as this without exploring it, when there is a possibility, nay perhaps a probability, that I may never again return to accomplish it, shews but little of that genuine spirit of discovery which contemns all danger and inconvenience when put in competition with its gratification! Upon that score of duty I might (it may be said) be forgiven, but must never again boast of a single spark of that ethereal fire with which the souls of Columbus and of Cook were wont to burn!—I am not indeed such a Quixote in discovery as this, although since I was able to read Robinson Crusoe, it has been within constant pursuit; but there is another reason remaining in aid of the first,—the debilitated state of my health, as well as of many others on the ship, and a lameness in both feet from incorrigible scorbutic ulcers, render me unable to go about any longer in boats, or to the masthead of the ship; both of which are absolutely necessary to any tolerable accuracy in this kind of surveying. I suppose it is unnecessary to state that the whole of this important part of our duty rests upon me: for Port Jackson, then, we now steered away, with a fresh and fair wind.[30]

Even with the arrival of the southeast trade winds, Flinders still could not bring himself to steer straight to Port Jackson. He continued to make several exploratory asides while they followed the west coast of New Holland homeward. These delays outraged Bell, who formally accused his captain of failing in his humane duty to his sick and dying men. While acknowledging Bell's concern, Flinders savaged the surgeon for questioning his authority and attempting "to raise yourself a character of Humanity, by putting a malignant stigma on mine . . . [H]ad the health of the people been the great object of my duty as it is yours," he thundered, "and had I been able to follow my own plan for their preservation, I should certainly have left them on shore in their native country, and not exposed them to the danger of the seas and enemies and to pernicious changes of climate, to all of which the execution of my orders makes it necessary."[31]

Nevertheless, the realization that eighteen of his crew were stretched out on hammocks below deck, some "almost without hope," eventually forced the man of feeling to sail quickly for Port Jackson. The *Investigator* reached Sydney Heads on June 9, 1803, but not before a succession of good men had been committed to the deep.

Matthew Flinders's achievements were epic by any standards. He circumnavigated the coast of Australia; he discovered that no strait or sea separated New Holland from New South Wales, thereby showing these to be part of the one continent; he named this continent Terra Australis, or Australia; and he substantially enhanced Cook's charts and discoveries on the northeast coast. Yet he cursed himself for failing to live up to his own expectations. The charting of the Torres Strait had been perfunctory, he'd had to abandon detailed surveys of large sections of the northern and northwest coasts, and he knew that the exacting botanist Robert Brown was disappointed at the paucity of their scientific finds. Finally, the death of excellent sailors like the quartermaster John Draper weighed heavily on his conscience.[32]

Later in 1803, while returning to England to report on this circumnavigation, Flinders was shipwrecked on Wreck Reef in the Coral Sea, and a few months later he found himself imprisoned on Mauritius by the island's overbearing French governor, General Charles-Mathieu-Isidore

Decaen. Three years later, still in lonely exile, he published an article in the *Philosophical Transactions of the Royal Society* that contained the first ever use of the term "barrier reefs." This denoted concealed coral reefs, separated from but growing parallel to the mainland. It seems likely that the deluge of obstacles he'd encountered over the previous five years inspired this most apt geographical coining.[33]

Flinders's litany of setbacks did not cease even after his return to England after a six-year imprisonment. In 1812–14, as he wrote up his long-delayed journal, *A Voyage to Terra Australis*, his life and career were once again assailed. His health had disintegrated to the point where he was now caught in a race against death to complete his manuscript. He won by a whisker, and it was in this great creative work, published the day before he died, that he gave a name for the first time to the massive assemblage of submarine coral obstacles he'd encountered along the east coast of Australia. He called them "the Great Barrier Reefs."

There is no doubt that this was the modern Reef's foundation document. Matthew Flinders, not James Cook, was the true European father of the Reef, because he was the first person to infer its unified existence and to conceive of it as a whole. *Terra Australis* brought together the romantic and scientific sides of his personality in a brilliant fusion. Using imagination and intellect, Flinders contended that the Reef was a continuous, interconnected work of nature and a coherent geographical region.

The book's summary revealed the reasoning processes, inferences, and speculation that conjured the Great Barrier Reef into being. Flinders argued that, though the portions of the outer reef sighted by the *Investigator* lay considerably eastward of those discovered by Campbell and Swain in the early 1790s, "there can be no doubt that they are connected." If this was so, he suggested, the reefs probably began as far south as Breaksea Spit, for "[it] is a coral reef, and a connexion under water, between it and the barrier, seems not improbable." At the point where the *Investigator* made its exit to the open sea, he thus estimated that the barrier reefs had already run continuously for some 350 miles.

After this, he admits the *Investigator* sailed too far northward to sight the Reef again until the Torres Strait. Yet since Cook had reported sailing in protected water all the way northward from Cape Tribulation, "I therefore assume . . . that with the exception of this [the Flinders En-

trance], and perhaps several small openings, our Barrier Reefs are connected with the Labyrinth of captain Cook; and that they reach to
Torres' Strait and to New Guinea . . . through fourteen degrees of latitude and 9 degrees of longitude; which is not to be equalled in any other
known part of the world." In short, Flinders was the first person to see
that coral stretched in a single connected train from Breaksea Spit almost to the coast of New Guinea, in what we now know to be the world's
largest reef.[34]

From this, Flinders also derived an appreciation of the geographical character and significance of what we today call the Barrier Reef
lagoon:

> An arm of the sea is inclosed between the barrier and the coast, which is
> at first 25 or 30 leagues wide; but is contracted to 20, abreast of Broad
> Sound, and to 9 leagues at Cape Grenville; from whence it seems to go
> on diminishing, till, a little beyond Cape Tribulation, reefs are found
> close to shore. Numerous islands lie scattered in this inclosed space; but
> so far as we are acquainted, there are no other coral banks in it than
> those by which some of the islands are surrounded; so that being shel-
> tered from the deep waves of the ocean, it is particularly well adapted to
> the purposes of a coasting trade. The reader will be struck with the anal-
> ogy which this arm of the sea presents to one in nearly the same latitude
> of the northern hemisphere. The Gulph of Florida is formed by the coast
> of America on the west, and by a great mass of islands and shoals on the
> east; which shoals are also of coral.[35]

Even so, he didn't gloss over the navigational challenges of coasting
through this great basin of water. The lagoon's depths were irregular,
unpredictable, and often excessively shallow, so that continual soundings
had to be made. Tides were more varied in incidence and intensity, as
well as larger and stronger, than most navigators would expect. Even
the wide Flinders Passage out of the Barrier required vigilance because
it contained "many small unconnected banks [of coral]." Flinders recommended that future sailors make a lagoon entry at Breaksea Spit and
an exit through his passage, but cautioned that "the commander who
proposes to make the experiment, must not . . . be one who throws his
ship's head round in a hurry, so soon as breakers are announced from

aloft; if he do not feel his nerves strong enough to thread the needle . . .
amongst the reefs, whilst he directs the steerage from the mast head,
I would strongly recommend him not to approach this part of New South
Wales."[36]

Flinders's cartographic overview of the Great Barrier Reef was matched
by his equally original, but often overlooked, series of scientific observations on coral reefs, which he made in the Torres Strait section of *A Voyage to Terra Australis*. Here he speculated, as a scientist rather than a
navigator, about what we would today call the geomorphological origins,
structures, and ecologies of coral reefs and cays. Exploring a small cay a
day's sailing from Murray Island had moved him to reflect:

> It seems to me, that when the animalcules which form the corals at the
> bottom of the ocean, cease to live, their structures adhere to each other,
> by virtue either of the glutinous remains within, or of some property in
> salt water; and the interstices being gradually filled up with sand and
> broken pieces of coral washed by the sea, which also adhere, a mass of
> rock is at length formed. Future races of these animalcules erect their
> habitations upon the rising bank, and die in their turn to increase, but
> principally to elevate, this monument of their wonderful labors. The
> care taken to work perpendicularly in the early stages, would mark a
> surprising instinct in these diminutive creatures. Their wall of coral,
> for the most part in situations where the winds are constant, being ar-
> rived at the surface, affords a shelter, to leeward of which their infant
> colonies may be safely sent forth; and to this their instinctive foresight
> it seems to be owing, that the windward side of a reef exposed to the
> open sea, is generally, if not always the highest part, and rises almost
> perpendicular, sometimes from the depth of 200, and perhaps many
> more fathoms.[37]

Flinders didn't yet know that reef-making corals need light to survive
and can grow only in relatively shallow waters, so he did not confront the
mystery of how the animalcules managed to build their "monuments"
within oceanic depths. Nevertheless he offered up a series of remarkably
shrewd observations about the environmental character and achievements of these tiny creatures. He inferred, for example, that they had to
be "constantly covered with water" to survive, "for they do not work,

except in holes upon the reef, beyond low-water mark; but the coral sand and other broken remnants thrown up by the sea, adhere to the rock, and form a solid mass with it, as high as the common tides reach."

Once these solid boulders were tossed clear of seawater, he noticed, they seemed to "lose their adhesive property" and to lie in loose jumbles that gradually developed into a "key" (cay) as sand gathered on the reef. Before long, in a series of stages, these cays gradually came to life: salt plants grew, soil formed, and birds carried over seeds of pandanus, coconut, and other shrubs and trees. Every gale piled up fresh mounds of sand, wood, broken trees, insects, and small creatures, until "last of all comes man to take possession."[38]

Flinders deduced, too, that the little cay in the Torres Strait that he'd called Halfway Island was well advanced in this "progressive" evolution. The lower part of the island—clear of the wash of even the highest spring tides—was still covered with half-evolved rock that displayed organic origins, such as "sand, coral, and shells . . . in a more or less perfect state of cohesion; small pieces of wood, pumice stone, and other extraneous bodies which chance had mixed with the calcerous substances when the cohesion began, were inclosed in the rock; and in some cases were still separable from it without much force."

The upper part of the cay, by contrast, was already covered in casuarinas and other shrubs and trees, which were in turn providing food for parrots, pigeons, and other birds, "to whose ancestors it is probable, the island was originally indebted for this vegetation."[39]

Not until thirty-two years later, when another young explorer visited the Cocos (Keeling) atolls in the Indian Ocean in search of answers to similar speculations about the origins and character of coral reefs and islands, would Matthew Flinders's luminous analysis be bettered. Significantly, that young man, Charles Darwin, had been reading Flinders's *Terra Australis* before he arrived, and he borrowed from it the arresting metaphor of coral reefs as vast "monuments" to the tiny animalcules that built them.

Neither was it a coincidence that both these young coral theorists, who shared Enlightenment and romantic sensibilities, were reading Milton's great romantic poem *Paradise Lost* at the time they made their reef observations. Flinders and Darwin also shared a belief that coral reefs and islands kindled mankind's deepest poetic and scientific faculties, for, as Darwin said, "such formations surely rank high among the wonderful objects of this world."[40]

But if any of Flinders's readers were inclined to see the Reef in a similarly romantic way, that inclination was soon to be dispelled by the harrowing testimonies of one Mrs. Eliza Fraser.

3

CAGE

Eliza Fraser's Hack Writer

THE STORY OF ELIZA FRASER'S ORDEAL at the hands of an Aboriginal clan at the southern end of the Great Barrier Reef resounds through white Australian history. Before this incident, most readers in Britain and Australia knew little or nothing about the Reef region, other than Hawkesworth's colorful account of Cook's battle with the Labyrinth. Over the years, Eliza Fraser's story has congealed into a core cultural myth, one of the few to be taken up by artists in other countries, and arguably as alive today as at the time of its inception. No one could possibly have foreseen its ramifications, which surely included the inclination on the part of many settlers to see Aboriginal peoples as violent, animalistic, and sexually predatory.

On September 27, 1837, John Curtis, court reporter for *The Times* of London, opened up a rival newspaper, the *Morning Advertiser*, to read in it a startling private letter. The letter had been sent several months earlier from the Liverpool Commissioner of Police, M. M. G. Dowling, to the current Lord Mayor of London, Sir Thomas Kelly, and it had warned the mayor of the suspected fraudulent conduct of "a person calling herself Mrs. Fraser."

Curtis already knew Mrs. Eliza Fraser, who was on the way to becoming a London celebrity. She was the widow of James Fraser, captain

Mrs Fraser in John Curtis's *Shipwreck of the Stirling Castle* (London: Virtue, 1838) (Fryer Library, University of Queensland Library)

of a trader called the *Stirling Castle*, shipwrecked a year earlier on a coral reef two hundred miles off the coast of northeast Australia. Toward the end of 1836, newspapers in both Britain and the Australian colonies had carried reports that Captain and Mrs. Fraser and a small party of sailors had, after their shipwreck, landed a longboat on Great Sandy Isle (which was renamed Fraser Island after Eliza) at the southernmost point of the Great Barrier Reef, only to be captured by tribes of fierce natives. These we now know to be several bands or subgroups within the Kabi Kabi language group—the Ngulungbara in the north, the Badtjala in the center, and the Dulingbara in the south.

After six weeks of living with the Badtjala, during which time Mrs. Fraser's husband and several other sailors died from maltreatment, she was rescued by a convict and returned to the Moreton Bay settlement, near present-day Brisbane. After recuperating for a while in Sydney, she embarked for England in early 1837 on a merchant ship owned by Captain Alexander Greene.

When she arrived in Liverpool on July 16, Mrs. Fraser approached the police to beg for relief from distress, and to ask for money to travel to London so as to take her horrific story directly to Lord Mayor Kelly. "But," Police Commissioner Dowling's subsequent letter informed Kelly, "on the second interview I had with her, an evident exaggeration of her sufferings while in captivity, caused a suspicion, and her relief was suspended till inquiries were made, when it turned out that she had married in Sydney . . . the master of the vessel in which she arrived here . . . who is a man in good circumstances, and who it now appears accompanied her to London . . . no doubt solely for the purpose of raising money by imposing on your Lordship and the public. Her husband, whose name is Greene, is the person who so warm-heartedly confirmed her statement before your Lordship."[1]

By the time he received this warning letter from Dowling, Kelly had already committed himself publicly to Mrs. Fraser's cause, so he delayed his reply in the hope of finding evidence to verify her claims. Annoyed at the snub, Dowling eventually leaked the letter to the *Morning Advertiser*, which embellished it with the headline: "MRS FRASER . . . whose extraordinary adventures among savages have lately excited sympathy, is now suspected of being an imposter." The radical *Morning Post* grabbed the opportunity to attack Lord Mayor Kelly, jeering that he'd failed to respond to the police warning because he'd "interested himself very warmly in her behalf."[2]

This was true. Facing an impending election, Kelly—a successful publisher of cheap books for the masses—had decided that Mrs. Fraser's shocking story of captivity and abuse by cannibals could be harnessed to his political cause. Taking up the plight of the brave widow would present him, Kelly reasoned, as a good-hearted philanthropist. When Mrs. Fraser and Captain Greene had approached him, Kelly did not question her account of how she'd been captured and tortured by savages, and left lame, half blinded, and destitute by the ordeal. Announcing that he'd "never heard anything so truly dreadful," Kelly urged the charitable ladies of London to give generously to his appeal for this "unfortunate" lady.[3]

Being also a chief magistrate, Kelly had decided to convene a Mayoral Court of Inquiry to give publicity to Mrs. Fraser's story. Favorable coverage of the event wouldn't, of course, come free. All early nineteenth-century British newspapers relied to some degree on income procured by small-scale bribery or blackmail—and *The Times* was no exception. Its editor and reporters always expected to be paid for "puffs" (positive mentions) or for "excisions" (the dropping of discreditable mentions). Money would need to change hands to ensure that John Curtis gave due prominence to Kelly's sympathetic interviews with Mrs. Fraser.[4]

And so it proved. Curtis's court reports presented a sparkling, real-life newspaper melodrama of cannibalism, imprisonment, murder, torture, and sexual violation—a story so affecting that the lord mayor's public subscription quickly topped the considerable sum of five hundred pounds. But now, to the great inconvenience of both Curtis and Kelly, the *Morning Advertiser*'s exposé of Mrs. Fraser threatened to turn their scoop into an embarrassing scandal.[5]

John Curtis was in an especially awkward position because he was also well on the way to completing a book about Mrs. Fraser's ordeal, for which he'd composed the juicy title *SHIPWRECK of the STIRLING CASTLE, . . . the Dreadful Sufferings of the Crew, . . . THE CRUEL MURDER OF CAPTAIN FRASER BY THE SAVAGES [and] . . . the Horrible Barbarity of the Cannibals Inflicted upon THE CAPTAIN'S WIDOW, Whose Unparalleled Sufferings Are Stated by Herself, and Corroborated by the Other Survivors*. News of Dowling's letter threatened to wreck the potential bestseller-in-progress, which Curtis hoped would appeal to a well-established vein of popular fascination with the sexual perversity, violence, and cannibalism of South Seas natives. Eliza Fraser's six-week "captivity"—the first entailing a woman castaway—needed only some amplification and reshaping to achieve a level of sensation suitable for a mass readership. Curtis's name, as the press mouthpiece of this wronged and fascinating lady, was already known through his *Times* court reports, which had generated a spate of imitative ballads, chapbooks, playbills, and merchandise from the grubby hacks in the London district of St. Giles and Seven Dials. His prospective book had everything going for it.[6]

But who was this John Curtis? The man remains a mystery. All we definitely know is that he worked as the court reporter of *The Times*. Given recent evidence of how interchangeable the names John and James were in Britain at that time, it is probable he was the same *Times* court reporter who wrote a decade earlier under the name of James Curtis. This James Curtis had also published a sizzling bestseller, based on his reports of the trial of William Corder, a well-to-do farmer's son accused of the grisly murder of a young mole catcher's daughter, Maria Marten. The public couldn't get enough of the seamy story. Curtis's *Authentic and Faithful History of the Mysterious Murder of Maria Marten* pioneered the genre of real-life courtroom drama, which some modern critics hail as the genesis of Truman Capote–style crime writing. Moreover, *Maria Marten*'s publisher was none other than Thomas Kelly, the same man who as lord mayor of London was now conducting the inquiry into Eliza Fraser's story.[7]

Possibly in order to avoid accusations of collusion, John Curtis had organized for his forthcoming book to be published not by Kelly himself, but by George Virtue, one of the mayor's friends. Virtue was another successful cheap-tract publisher, who worked almost next door to Kelly on Paternoster Row, and who served with him for the same London shire on the Court of Common Council, an elected body of the Corporation of London.[8]

Even if James and John Curtis were not the same man, the latter's decision to use the former's bestseller as a model for the Eliza Fraser book must have been irresistible. *Maria Marten* had been a spectacular financial and popular success. Eliza Fraser's story—suitably embellished—contained similarly sensational ingredients of torture, sex, and violence. In the introduction to his book, John stressed that he'd adopted a true-life, courtroom-drama approach that presented testimonies "from the lips of such of the survivors as we could have access to." This, he claimed, had enabled him to eclipse all existing accounts of the affair in newspapers and chapbooks, which offered "a mere *epitome*." By contrast, "it will be our object to narrate and arrange [the testimonies] in the chain of melancholy recital."[9]

Unbeknownst to himself, Curtis was actually doing far more than this. He was writing the book that would become the primary source for all subsequent retellings of the Eliza Fraser story, up to our own times. It is for this reason that we need to explore the book's origin, structure, and content in detail.

Curtis's introduction also offered other enticements to the reader. These included harrowing accounts of head-hunting and cannibalism—practices familiar from South Seas castaway memoirs like that of Alexander

Selkirk, Defoe's real-life model for Robinson Crusoe—as well as elements from Native American "captivity" tales. Curtis further promised all the attractions of romantic travelogues—dramatic engravings, sentimental poems, intimate personal letters, and spiritual homilies—and these would be seasoned with self-improving doses of scientific, topographic, navigational, and ethnographic information. "Useful knowledge" of this sort provided both the pleasures of entertainment and the certainties of factual authority; Curtis's observations would reveal "the manners and customs of the barbarians, among whom the sufferers were cast." All this "unquestionable corroborative testimony" would dispel any doubts "that human nature . . . could have borne up under tortures so numerous and enduring, and insults so diabolical."[10]

Curtis assured readers that his narrative was based primarily on the court testimonies of Mrs. Fraser and her supportive witness, John Baxter—the nephew of her late husband and second officer of the *Stirling Castle*. Even before this, however, the journalist had managed to obtain some rich copy directly from Captain Greene and Eliza themselves, whom he'd befriended as soon as they arrived in London. The earlier versions Mrs. Fraser had given to the commandant of the Moreton Bay settlement and then to Sydney newspapers were rather too reticent.

True, even at that early point she'd complained of having experienced physical and mental hardships while living for six weeks with small bands of Aborigines on Great Sandy Isle: they'd stolen her clothes, forced her into backbreaking labor, and cruelly hastened the deaths of her husband and of First Officer Charles Brown. However, she'd said nothing to the Moreton Bay commandant about murder, torture, or acts of cannibalism. She'd talked only of the harsh treatment meted out by the Aborigines on whom she'd relied for survival. Her ravaged condition appeared to have come about from foraging for scarce food for those six weeks without protection from the searing heat and torrential rain.[11]

But Curtis's book told a very different story. His fictive talents were everywhere to be seen. He provided fervid descriptions of Eliza's anguish, revelations of her maternal and Christian virtue, hints of diabolical sexual violations too disgusting to specify, and lingering details of bestial native cruelties and cannibal practices. A true literary craftsman, he also inserted a series of tactical digressions designed to ramp up the tension or to gloss over inconsistencies in Eliza's story. Wherever possible, he omitted any statements by court witnesses that might have the undesirable effect of humanizing her captors.

In short, Eliza Fraser was presented as the epitome of early Victorian womanhood, a lady who had exhibited inspirational qualities from the opening moment of the tragedy, on May 21, 1836, when the *Stirling Castle* hit one of the Swain reefs in deep water at the southeastern entrance to the Great Barrier Reef.

> The solemnity and terrors of that awful night, were heightened in a great degree by peals of thunder and flashes of vivid lightning, such as have never been heard or beheld in our latitude; the elements above seem to have confederated together with those beneath, to strike alarm and dismay into the minds of the benighted and shipwrecked captain and his desponding and exhausted crew; and were imagination to be expended to its utmost bounds, it could form no adequate idea what must have been the sensations of one person on board the wreck,—a woman, a doating and affectionate wife, one, who being influenced by conjugal fidelity, and anxiety for the health and welfare of her husband, had left her country, children and friends, to console him in the hour of sickness and exhaustion, from a consciousness, that while performing the duties which the law of connubiality enjoins, she had no reason to dread the terrors of the mighty sea.[12]

Eliza's husband also needed some sanitizing. Captain Fraser's record of a previous Barrier Reef shipwreck, a chronic inability to retain crewmen, and a paranoid terror of cannibals was not helpful. Rather than portray him as already ill and half crazed with fear on first meeting Aborigines on the Great Sandy Isle beach, Curtis declared that he had from the outset "been a marked man, [who] underwent more suffering and experienced more contumely" from the savages than any other of the castaways. Reckless of the consequences, however, the captain had chivalrously "interfered in behalf of his wife, when he beheld her subjected to diabolical insults."[13]

Fraser's refusal to do the "bidding" of the savages, claimed Curtis, caused them to contrive "ingenious and horrid modes of torture" that culminated in a cowardly murder. A spear, hurled from behind, "struck him near the shoulder blade, and passing through his body, came out at his breast." His brave wife "darted from her hiding place and exclaimed,

'O Jesus of Nazareth! can I stand this?'" whereupon her husband, expiring in a welter of blood, whispered, "'O Eliza! I am gone!'" "These were," said Curtis, "the last words uttered by the unfortunate victim of barbarian vengeance." An accompanying illustration of the fleeing captain being impaled through the back completed one of the book's great set pieces. To ensure that the scene achieved its intended effect, though, Curtis also needed to suppress Eliza's surprising admission to Mayor Kelly's court: "I don't think it was their intention to kill my husband; I believe the man who cast the spear merely intended to wound him." She implied that Captain Fraser's already wasted body had succumbed to the shock of the spearing.[14]

Similar omissions were needed for Curtis to make a convincing case that two other castaway crewmen had been murdered. He presented the death of First Officer Charles Brown as a stock Native American–style atrocity: "Mr Brown was inhumanly tied to a stake, and a *slow fire* being placed under him, his body, after the most excruciating sufferings, was reduced to ashes." The slowness of the fire was to ensure "that their [the savages'] joy might be enhanced at the writhing of their victims." John

The murder of Captain Fraser as depicted in John Curtis's *Shipwreck of the Stirling Castle* (London: Virtue, 1838) (Fryer Library, University of Queensland Library)

Baxter and Eliza Fraser admitted in court that neither had been present at the man's demise, but this didn't stop Curtis from citing them as first-hand witnesses.[15]

He also included a claim that Baxter had found fragments of bone on the beach that he identified by scraps of clothing as coming from a young sailor named James Major. Although Major appears to have died a natural death and rolled into his own campfire, Baxter's evidence inspired Curtis to write that "the natives had placed [Major's] head on a fire, which consumed the thorax, and descended obliquely to a part of the left side of the abdomen, when it appeared to have satiated its vengeance, or perhaps its flame was extinguished by the gushing of the heart's blood of the victim." Curtis's earlier *Times* report had been equally imaginative: "the savages," he wrote, "set to work and by means of sharpened shells severed the head from the body with frightful lacerations. They then ate parts of the body, and preserved the head with gums of extraordinary efficacy and affixed it as a figure bust to one of their canoes."[16]

Curtis regretfully had to leave these flesh-eating details out of the book, because Baxter suddenly forgot them when testifying to the mayoral court. It was a pity: confirmed instances of cannibalism would have proved that these natives were crueler than "wild beasts." But details of the inquiry, which was held at the same time that Curtis was writing his book, were being published in many newspapers, forcing him to curb his fictional impulses. He had to content himself instead with a few rather unsatisfactory generalizations, such as his definition of a "corrobery" as a savage dance "round a miserable captive, whose flesh they would presently greedily devour." He also dug up some unsupported accusations of cannibalism made by a naval surgeon during the 1820s, which were directed at Aboriginal clans living one thousand miles from Great Sandy Isle. Mrs. Fraser's savage captors, Curtis implied, had shared those Aborigines' love of eating human flesh "in a manner the most revolting." They would consume enemies slain in war, as well as any Europeans they could capture, and—when especially hungry—even eat their own children.[17]

Curtis felt himself on stronger ground when bringing to light new instances of hardships endured by Eliza Fraser and her associates. The most startling of his scoops was the revelation that she'd been forced to give birth to a premature baby in the ankle-deep bilge water of the long-boat. The baby had died, "after gasping a few times," and the body was tossed overboard wrapped in a piece of shirt. Curtis quoted a hymn by his clergyman friend, Reverend G. C. "Boatswain" Smith, which compared the dead babe to one even more famous: "In 'swaddling clothes' wrapt midst infernal commotion, / To sink and to die on life's accurst ocean."[18]

Curtis also disclosed fresh details of suffering during Eliza's "worse than satanic bondage" ashore, especially after she was assigned to the care of a band of native women. They had evidently howled "with derision and mirth" at her naked, sunburned body and then, as if she were an animal in a cage, pelted her with wet sand that caused "excoriation of the skin . . . excruciating almost beyond endurance." This entertainment triggered a succession of worsening tortures that included starving her of food, forcing her to carry heavy logs, making her nurse a lice-infested infant, beating her with clubs, stabbing her with spears, and burning her legs with lighted brands. Eliza's ordeals, Curtis speculated, must have been motivated by the native women's jealousy, which was aroused "because attentions of a diabolical nature were paid her by the men." As for the men themselves, their "sport" supposedly consisted of throwing lighted resinous bark on the sleeping castaways, or tossing them overboard from canoes at sea, "for the purpose of exulting in their struggles to save themselves from drowning."[19]

After relating this litany of horror, Curtis suddenly thought it prudent to reassure readers that Mrs. Fraser and Mr. Baxter had in no way "over-coloured their statements in respect of the suffering which they endured." Pages later, a more urgent footnote on the same subject temporarily displaced the text, spilling over into several additional pages of tiny print.

These interventions marked the moment, in September 1837, when Commissioner Dowling's letter hit the news. Curtis admitted in the footnote that he'd been suddenly confronted in the middle of writing with a flurry of newspaper accusations that Eliza Fraser was "an ingenious impostor," a "base fabricator," and a purveyor of lies "to gull the benevolent." He felt bound, therefore, "to deviate from the track . . . originally marked out" in order to defend his heroine against such "cruel and un-English" abuse.[20]

Here Curtis abandoned all pretense of being a dispassionate court reporter and flung himself into the role of defending attorney. Most of his emergency footnote was given over to reproducing a letter just ob-

tained by Lord Mayor Kelly from Lieutenant Charles Otter of the Moreton Bay settlement. A year earlier, Otter had masterminded Eliza Fraser's
rescue by conscripting an Irish convict familiar with the Kabi Kabi
people to persuade the natives to hand her over. Otter described his first
sight of the rescued lady: "You never saw such an object . . . Although
only thirty-eight years of age, she looked like an old woman of seventy,
perfectly black, and dreadfully crippled from the sufferings she had undergone. . . . She was a mere skeleton, the skin literally hanging up on
her bones, while her legs were a mass of sores, where the savages had
tortured her with firebrands."[21]

But was this a sufficiently harrowing picture to persuade Curtis's
readers that Mrs. Fraser was innocent of subsequent exaggeration and
financial fraud? On further reflection he decided that his case needed
strengthening, so he introduced a whole new chapter in which to defend
Eliza. In doing so he compared himself to a navigator faced with an unexpected reef, "which called into requisition all our nautical experience,
and forced into active service all the skill in seamanship which we possessed." Moreover, colonial newspapers were now also joining in the
clamor. They accused Eliza of "inexcusable ingratitude" in concealing the
generous financial help given by the Australian public. One Sydney newspaper pointed to a marked disparity between her original and recent London testimonies, stating that the new version of her suffering was "greatly
overcharged."[22]

Experience as a court reporter had evidently taught Curtis a thing
or two about how to undermine a prosecution case. He decided on a
threefold strategy. First he reminded his readers of the "gallant" Lieutenant Otter's proof that Mrs. Fraser's ordeal had "been of a very extraordinary kind." In fact, Curtis hinted that delicacy had forced him to
understate it: "we have in our possession facts connected with the brutal
treatment of this helpless woman, (and could produce a living witness
who would verify them on oath,) which, if we dared to publish, would
excite an involuntary shudder of horror and disgust in every well-regulated
mind."[23]

Next he endorsed a shrewd tactical move just made by Kelly. His
lordship had remembered that Mrs. Fraser's London subscription had all
along been intended for her three grieving and indigent children in the
Orkneys, who were being looked after by a local clergyman. In one
stroke this transformed Kelly's act of self-interested credulity into a

"meritorious and praiseworthy" piece of public philanthropy intended to benefit "destitute orphans." It also justified Eliza's desire to raise more money in London, since she'd wanted it only to alleviate "the destitute condition of her three poor children, the legitimate offspring of a brave and unfortunate man." To underline the point, Curtis added a poignant letter, supposedly written by Eliza's fifteen-year-old daughter, Jane, explaining how her "dear mother" had gone to London to obtain money on behalf of her and her two brothers.[24]

The third element of Curtis's strategy entailed some delicate maneuvering, because he had to justify Eliza's failure to disclose her marriage to Captain Greene in Sydney, by which she'd ceased to be a destitute widow. He wisely decided to gloss over the issue of her financial assets and instead treat the matter as a minor breach of social convention. It was the case, he conceded, that she'd married Captain Greene relatively soon after the death of her husband, but who among his readers could predict or wish to control the ways of true love? "Here she was in Sydney, in a state bordering upon utter destitution. She became acquainted with Capt. Greene, a gentleman well-known and highly respected there . . . Perhaps he first viewed her, as did hundreds of others, as an object of commiseration; and at length pity gave way to a platonic affection, which ripened into a more tender sensibility."

Both the Captain and his wife regretted the "indiscretion" of failing to admit their married state when they reached England. They had kept silent on this subject purely for technical legal reasons: there'd never been any intention to deceive. Curtis beseeched his "readers to forget her error, which at most is a venial one . . . She is fully aware that she has sinned against strict etiquette, and been guilty of an indiscreet secrecy; and we are ready to admit these facts; but without her knowledge, we have attempted an apology."[25]

Though well satisfied, as he wrote, with "the manner" of his reply to Mrs. Fraser's critics, Curtis still had a further reef to navigate. Late in the lord mayor's inquiry, a surprise witness had appeared in the form of "Big Bob" Darge, one of three sailor castaways from the *Stirling Castle* on the island who'd been attached to a different group of natives from Eliza Fraser (probably the Ngulungbara in the north of the island). These three—Robert Darge, Joseph Corralis, and Henry Youlden—had walked to the mainland to relay the news of Eliza's situation to Lieutenant Otter.[26]

Darge was proving to be an uncooperative witness. He was a tough sailor with no particular liking for his Aboriginal hosts, but his testimony contradicted the bleak picture presented by Eliza Fraser and John Baxter. Darge complained mainly about the rigors of the coastal environment, the scarcity of food, and the hard work needed to survive. But he didn't pretend to have been worse off than the Kabi Kabi themselves: "it was the winter of that part of the world when we were with the natives, and they had a great deal to do very often to manage to live." Furthermore, he openly admired the natives' skill in fishing and hunting and declared their main food staple, the *bungwa* root, to be "delicious."

Kelly fired a succession of leading questions at the burly sailor in an effort to corroborate Eliza's indictment of the "savages," but Darge refused to oblige. He denied that the native males were brutal toward their females, denied that they lacked affection for their children, and denied that the men had taken pleasure in tormenting him. He even denied that they'd had any intention to kill him. And his admission that some of the natives hated white men because of the violence inflicted on them by local white soldiers hardly helped Mrs. Fraser's case. Above all, Darge flatly refuted "that any of the tribes I was among ate human flesh."

Exasperated, the lord mayor asked: "You don't seem to think these natives such desperate savages as Mrs. Fraser and [Mr.] Baxter considered them to be?"

"I was certainly treated with great roughness," Darge replied, "but I don't think they would kill intentionally." In fact, he said, it was only thanks to the bush skills of voluntary Aboriginal guides that he and the other two sailors had reached the Moreton Bay settlement.[27]

Curtis couldn't exclude Darge's testimony because it was already circulating in British and colonial newspapers, but he decided to prevent readers of his book seeing a verbatim version of the interrogation. Instead he provided a shrewdly edited interpretation of the sailor's evidence. Like a skilled defense lawyer, he used Darge's responses as a way to cast doubt on his objectivity, depicting him as a man of mutinous disposition, and he also, wherever possible, twisted Darge's words so as to endorse Eliza's claims.[28]

Curtis attributed the natives' relatively humane treatment of Darge and his companions to sheer chance; he contended that the three castaways were lucky enough to fall into the hands of a more benign "tribe" than had Mrs. Fraser, whose cannibal captors represented "the zero of civilization." Darge, "being naturally an abler-bodied man," was "evidently considered" "more valuable than some of the less muscular portion of the captives." Furthermore, Curtis suggested, the tribe chose to mitigate their natural cruelty because Corralis, one of the three sailors, also happened to be black. He concluded his case with the news that Big Bob Darge was about to undertake a fresh voyage to "a remote quarter of the earth," even though "it was quite apparent that the health of the poor fellow had been greatly impaired."[29]

This rather lame last sentence marked the surprising termination of the book—or at least, as Curtis explained in a footnote, "of what in strictness may be denominated the 'Narrative' of the *Shipwreck of the Stirling Castle*." It seemed an oddly feeble conclusion to his long and impassioned defense, but in fact he wasn't finished yet; he'd thought of an ingenious new way to advance his case. In an adjacent footnote he suddenly introduced his readers "to . . . another catastrophe, which in many respects is more appalling in its details than that which has preceded it."

When Eliza Fraser had originally arrived in Sydney after her ordeal, he explained, she'd found the citizens there already in a state of outrage at the behavior of the Barrier Reef savages, "owing to several other wrecks which had recently taken place in the vicinity of Torres Straits, particularly that of the *Charles Eaton*, whose captain and crew, as well as every person on board, were murdered, save a lad of the name of Ireland, and a child named D'Oyley, the son of a [military] captain, whose life was doubtless spared in consequence of the sagacity of the youth who was his companion and protector."[30]

Curtis therefore felt bound, he said, to provide his English readers with additional insights into the cannibal problem offered by the *Charles Eaton* story. "After giving a narrative of this dreadful calamity, it will become our duty to give extracts from documents connected with *both histories*, historical quotations, and other interesting communications, together with such original remarks and reflections as upon review may be deemed necessary."

To further link the two stories, he then inserted another hymn by his friend the Reverend "Boatswain" Smith, celebrating the "providential escape" of the survivors from both shipwrecks—a signal that Curtis would henceforth treat them in tandem. The title of this second part of the book also echoed its predecessor: *Narrative of the Melancholy Wreck of the CHARLES EATON, on One of the Barrier Reefs in the Torres Straits: with an Account of THE MASSACRE of the Captain, Passengers, and Crew; and of the Providential Rescue of John Ireland, aged 16, and WM. D'Oyley, aged 3, from the Savages.*

Despite the fact that the *Charles Eaton* had been shipwrecked two years before the *Stirling Castle*, news of the fate of its crew and passengers had leaked out only in dribs and drabs—in this sense it was still breaking news. Moreover, it offered an even more sensational set of horrors than Eliza's captivity.

Among the victims of the *Charles Eaton* were Captain and Mrs. D'Oyley, the parents of one of the rescued boys. The D'Oyleys belonged to East Indian military and merchant dynasties with substantial political clout, and Curtis was a sharp enough journalist to know that political clout could usefully translate into publicity. Charlotte D'Oyley's brother, a wealthy Stockton merchant named William Bayley, had lobbied relentlessly to find out what happened to the ship. Having formed "a kind of social compact,—a society of mourners weeping for their kindred," Bayley had bombarded the Colonial Office and the Admiralty with requests for news and action. That Bayley was a close friend of Sir John Barrow, a renowned Admiralty Secretary and wire-puller, also helped the cause of the grieving group. On Barrow's advice, Bayley sent letters to Mr. Stephens of the Admiralty Office, Sir George Gray of the India Office, John Cowan—who had recently replaced Thomas Kelly as lord mayor of London—and to the head of the Colonial Office, Lord Glenelg. All replied deferentially.[31]

Reports of sightings of the *Charles Eaton*'s wreckage had first trickled back to England in early 1835, two years before Eliza Fraser was rescued. Later in 1835 some survivors who'd escaped the shipwreck in a cutter reached Batavia. There they filed details of how, on August 15, 1834, the ship had been pounded to pieces on the "Detached Reef" near the entrance to the Torres Strait. Then, in February 1836, a Canton (Guangzhou) newspaper published a tantalizing letter written by a trading captain, William Carr. He claimed to have made fleeting contact on Mer (Murray Island) in the Torres Strait with a youthful *Charles Eaton* survivor, who'd refused to be enticed aboard his vessel. Carr said he'd also glimpsed a small white child on the shore, as well as signs of further survivors, but he'd not dared to make a rescue attempt. Carr's later testimony at another mayor's court in November 1836, attended by William

Bayley, helped precipitate the government's commissioning of two rescue ships to investigate his report, the HMS *Tigris* from India and HMS *Isabella* from Sydney.

In that same month of November 1836, Captain M. Lewis of the *Isabella* returned to Sydney with the only two *Charles Eaton* survivors, John Ireland and William D'Oyley, who after their long sojourn in the Torres Strait were now aged nineteen and four respectively. Both were in good health, but scarcely able to speak English. Ireland nevertheless managed to stumble out a horrifying tale. After landing on Boydang Island in the Torres Strait with two rafts of shipwrecked survivors, both boys narrowly escaped a subsequent massacre. Seventeen passengers and crew were clubbed to death, decapitated, partially eaten, and had their skulls bound into a mask. Captain Lewis had found and brought this gruesome trophy back to Sydney, where it generated waves of outrage. Around the same time, John Ireland met and spoke to Eliza Fraser in Sydney, but was decisively upstaged by her in press interviews. Ireland eventually reached London from Sydney in August 1837 and immediately presented his tale to another mayoral inquiry. A print version of the story appeared the following year, under the title of *The Shipwrecked Orphans*.[32]

Curtis had originally intended to offer only a short sketch of the fate of the *Charles Eaton*. Anything more ambitious was impossible because he lacked access to the firsthand oral testimonies that Eliza Fraser had given him. Neither could he waste time researching the new story; he needed to publish Eliza's saga before the public lost interest. It looked as if Curtis would have to rely on brief and irregular reports from colonial newspapers.

Sometime in late 1837, though, the journalist's luck changed. He stumbled on an obscure pamphlet from that year called *Narrative of the Melancholy Shipwreck of the Ship Charles Eaton*. Written by the Reverend Thomas Wemyss, a clergyman from Stockton in the north of England, it contained a well-documented, up-to-date account of the shipwreck, the massacre on Boydang Island, the voyages of the rescue ships, and the recovery of the two surviving boys from Mer. This was almost too good to be true: Curtis didn't know it, but the author was a personal friend of William Bayley, who had collaborated in producing the book. Wemyss had been given full access to Bayley's correspondence, including documents from the Colonial Office and Admiralty, private letters between

Charlotte D'Oyley and her two older boys, and poignant notes from a clergyman whose son had perished in the massacre.[33]

Wemyss supplemented his gripping account with information drawn from missionaries, newspaper interviews with John Ireland, and a recent Sydney pamphlet by William Brockett, one of the officers who'd sailed on the *Isabella*'s rescue mission to the Torres Strait.[34]

Best of all, as far as Curtis was concerned, the provincial obscurity of Wemyss's pamphlet meant he could steal its contents with impunity. Pressmen of the day needed to be skilled in the art of plagiarism, operating as they did in a free-for-all print economy, and Curtis was among the best of them. His boldness was breathtaking. Few other *Times* journalists would have had the bravado or ability to plunder an entire book, including its full title, so near to the time and place of its original publication.

Curtis opened his plagiarized story with Wemyss's moving reflection on the vicissitudes of a sailor's life. The piece was set in quotation marks, but attributed only to "a modern author," so that the source could not be traced. After this, he reproduced the entire pamphlet as his own, taking care at the same time to introduce some strategic alterations. At several points he simply rearranged the clergyman's original structure. He changed occasional words and phrases, sometimes for concealment, sometimes for greater effect. He inserted a few chunks of his own supercharged prose here and there, especially at points where he felt Wemyss had been too restrained. And he excised, when necessary, passages wherein the clergyman appeared overly tolerant of the Torres Strait natives.

Another of Curtis's tricks was to imply that information lifted from Wemyss's documented sources was given to him in face-to-face interviews. He even had the effrontery to insert a passing note of thanks to "*Mr. Wemyss*, a gentleman to whom we beg to express our high obligations for the occasional assistance he has rendered us."[35]

Not all of this word surgery was purely for disguise. Curtis wanted to weld this new piece to the story of Eliza Fraser so as to produce something like a super-text with a unified structure, style, and set of aims. To achieve this, he inserted strategic cross-references between the first and second stories, most of them referring to observations and claims made by Eliza and Baxter, and designed to cover weaknesses in her account by fortifying them with the meatier material of the *Charles Eaton* story.

With this twinning, Curtis was able to extend and transform the thrust of the overall work. He could now suggest that the castaways of the *Stirling Castle* and the *Charles Eaton* were victims of a common system of terror that threatened British merchant ships along an entire region of land and sea. His book, which he'd begun as a specific defense of Eliza Fraser and the lord mayor, could be repositioned as a far-reaching exposé that demanded government action on the appalling dangers confronting Britons within the Great Barrier Reef. By lucky chance, Great Sandy Isle, the site of Eliza Fraser's ordeal, and Boydang Island, where the massacre of the *Charles Eaton* castaways had occurred, marked the extreme southern and northern tips of the Great Barrier Reef. Scores of shipwrecked castaways generated within this vast minefield of coral, Curtis suggested, were being routinely captured, imprisoned, and subjected to fates worse than drowning. They became either the slaves or food of savage cannibals who shared proclivities to bestiality and violence. With his book, Curtis had produced a searing double indictment "of the coast, and the natives which inhabit it."

He demanded that something drastic be done about the Torres Strait in particular.

> The Straits of Torres . . . seem really as if they were destined to be the terror of navigators. This arises from the extreme difficulty of steering through that perilous passage, the irregular courses of the tides, the sudden manner in which storms and hurricanes arise, and the numerous shoals which are scattered in this vast expanse of water seem to bid defiance to nautical skill, and the steadiest caution. To detail the various wrecks which have happened there, that have come to our knowledge, would fill a large folio, and many a vessel has, doubtless, foundered, and been swallowed up in that insatiate gulf, of the particulars of which the world will ever remain ignorant. It is not unlikely that the sanguinary character of the natives, who massacre the survivors who fall into their hands, is the most plausible reason which can be assigned why the fates of many other hopeless vessels are never made known.[36]

While echoing Wemyss's pleas for the Admiralty to undertake urgent navigational surveys of the Torres Strait and Barrier Reef, Curtis's real agenda was that action be taken against the native inhabitants. Building on Wemyss again, he cited four possible solutions to the cannibal problem. The government could send a force to take possession of the islands "and then exterminate the whole of the inhabitants"; it could forcibly transfer all the islanders "to the coast of New Holland, and abandon them to their own natural resources on that vast continent"; it could "subjugate the inhabitants," "make them tributary," and try "to civilize and improve them"; or, finally, it could soften their sensibilities through the introduction of the gospel.[37]

Reverend Wemyss, a liberal man of God, had balked at genocide and favored the last course, dismissing the other three options as inhumane and unjust. He even suggested several times that the natives might have been goaded into their violence by previous bad experiences at the hands of Europeans. But his fiery plagiarist could not agree that the answer to this cannibal problem lay with missionaries, who were too soft and un-worldly to cope with Reef natives. Instead, since "these islands are prob-ably destined at no distant day to be important specks in the map of British territory," Curtis urged the setting up of "a Civilization Society." Organized groups of white settlers armed with mechanical and agricul-tural knowledge, and supported by soldiers, could domesticate the natives and "prepare their minds for the reception of gospel instruction."[38]

In order to ensure that his role of stern imperial prophet was in no way undermined by the facts, Curtis took care to include only the most cursory acknowledgment of how the two young *Charles Eaton* survivors had been rescued from their original captors by a kindly senior man from Murray Island called "Old Duppa," who had then adopted them into the clan. Old Duppa had treated John Ireland, now known as Wak, as his own son. The boy was given a tomahawk, a bow, and a sixty-foot canoe imported from New Guinea to fulfill his fishing and hunting needs, and he was granted his own parcel of land for cultivating yams, bananas, and coconuts. According to the boys' rescuer, Captain Lewis, another adoptive Mer family showered little William D'Oyley with love. Known as Uass, he forgot his European mother completely and attached himself devotedly to his new parents. Sturdy and browned by the sun, the boy spoke only the native language and cried bitterly when taken from his Mer home.[39]

By contrast, Curtis delighted in the gothic possibilities of the *Charles Eaton* story. Two incidents in particular, which he cited again and again, became keystones of his text. The first was John Ireland's account of the massacre of his fellow crewmen after they'd drifted in their half-submerged raft to Boydang Island. Curtis laced Wemyss's rather subdued version with gorier details from the *Sydney Times* of November 19, 1836:

> When they [those on Ireland's raft] first landed, the natives, with that lurking treachery which appears inherent in their natures, by their gestures and deportment appeared to be friendly . . . The hungry and fatigued crew sat themselves down, and several of them fell asleep on the spot where they halted,—the commencement of the sleep of death! No sooner had the dastardly ruffians discovered that their victims were asleep, than a multitude fell upon them, and commenced the work of general slaughter; spears, knives, and waddies being called into active requisition, for the purpose of destruction. Having deprived the poor fellows of life, they next cut off their several heads, and then joined in a corrobery around the bleeding victims, . . . uttering wild and discordant yells of joy . . .
>
> *Ireland* states that the savages . . . feasted upon the eyes and cheeks of the persons massacred by them belonging to the *Charles Eaton*. It is stated that these rude barbarians are induced to this horrible custom, from a belief *that such conduct will increase in them a more intense desire after the blood of white men.*[40]

The second ghoulish incident was the making of the skull mask, which appealed because of its diabolical nature. After taking the two boys aboard at Mer, the *Isabella* had set out to find and punish the perpetrators of the massacre. On reaching the cannibals' base at Aureed Island, Captain Lewis found it deserted, but discovered what he called a "Golgotha," or a "Temple of Skulls." It contained an ornate mask made up of forty-five human skulls, seventeen of which later proved to be from the *Charles Eaton* castaways. Curtis, unaware of the exact number of skulls, supplied his own words:

> The party having collected together, it was determined to enter the grotesque building, if an excavated and infernal den is worthy of such an appellation. They had not entered a moment, before the party in advance were horror-struck at beholding a large figure composed of tortoise-shells, to which were appended the skulls of several human beings. They were fixed to it by pieces of European rope, and some of the bones exhibited marks of violence, such as might have been inflicted by

the force of the massive waddies, sometimes used by the natives in the work of death. . . . There can be no doubt, we think, but that these were the relics of the mortal remains of some of our countrymen, who have been wrecked in these terrible straits.[41]

Curtis declared himself well pleased with his magnum opus. In a short conclusion he summarized its achievements, among which he listed his account of "the manners and customs of the aborigines, and the natural history of the islands in which their habitations are located," and also his moral history, which "exhibits not only a detail of the barbarity of the heathen, but also the benevolence of the Christian." He had, he said, provided a range of expert opinions, including his own, on how to control and civilize the "barbarous natives" of that part of the world. The last page of his book carried an engraving of the cannibal mask found in the Temple of Skulls, an "emblem of barbarity" that the resourceful Curtis had worked so long and hard to prove.[42]

The book quickly passed through several editions, yet its author could not have imagined in his wildest dreams how influential it would ultimately become. John Curtis, *Times* court reporter, never intended to

The mask of skulls in John Curtis's *Shipwreck of the Stirling Castle*
(London: Virtue, 1838) (National Library of Australia)

produce one of the foundation texts of British colonial and postcolonial culture, let alone a book that spawned a legend that still flourishes in the twenty-first century. He was in many ways a typical predatory journalist of his day: he simply wanted to concoct and sell a sentimental, racist, and sensationalist "true life story."

Over time, the story has been recast to suit shifting concerns. In the hands of the great Australian nationalist painter Sidney Nolan, the Eliza Fraser story became a saga about a convict outsider and a ravished lady. The convict, as much as Eliza herself, became an embodiment of Australian nativism pitched against an archaic British empire. Nobel Prize–winning author Patrick White saw it as a parable about the encounter between a "civilized" Englishwoman and the elemental forces of a harsh landscape and its native peoples. A film by Tim Burstall and a musical collaboration by Peter Sculthorpe and Barbara Blackman added new nationalist inflections again. What all these permutations had in common, though, was indifference or hostility toward the Aboriginal people who had given Eliza succor so that she lived to tell the tale.

Among the earliest critics of this bias were Queensland historians Raymond Evans and Jan Walker, who in 1977 drew on archival and anthropological information to contest the case that the Kabi Kabi were motivated by cruelty. In the wake of Australia's bicentennial year in 1988, other Europeans began to rethink the story, in a wider process of recognizing that white Australians had repressed much of their early history of Aboriginal dispossession, murder, and cultural destruction. Gillian Coote's documentary film *Island of Lies*, released in 1991, was a notable example of this recognition, featuring interviews with a Badtjala woman and a long-time Fraser Island settler, both of whom believed that the story was responsible for spreading fabrications about Badtjala cannibalism for the purpose of financial gain.

The same year saw an even more remarkable work, exhibited in Sydney, by Badtjala artist Fiona Foley. *By Land and Sea I Leave Ephemeral Spirit* was a haunting sequence of paintings and installations that reworked several images from Curtis's book. One showed Eliza Fraser juxtaposed with votive candles, suggesting the sacral nature of her story for European Australians; another depicted her snagged in a rat trap, symbolizing the verminous role that the legend has played in the lives of the Kabi Kabi and in those of Aboriginal peoples generally.[43]

Ironically, Eliza Fraser herself did not benefit from Curtis's nimble pen. There are unconfirmed claims that she was eventually forced into the ignominy of performing her story for a few coins in England's fairgrounds. If so, she ended up as much a victim as the Kabi Kabi she slandered. As Fiona Foley's use of Curtis's book shows, his text remains the vehicle of the most toxic version of the myth, more than 150 years after it was first published. Such is the enduring influence of that sly London hack, who has much history to answer for.

4

BASTION

Joseph Jukes's Epiphanies

THANKS IN PART TO JOHN CURTIS, cannibals were much on the mind of the British Admiralty in the spring of 1842, when it gave orders to the naval corvette the *Fly* to survey the northern end of the Great Barrier Reef and the surrounding waters and reefs of the Torres Strait. The Admiralty wanted particular attention paid to this area because so many British vessels trading in the South Seas or with India had come to grief trying to navigate the uncharted coral reefs and the strait's perilous narrow entrances. Since the wreck in 1791 of the ship sent in search of the *Bounty*'s mutineers (the *Pandora*, under Captain Edwards), more than twenty further losses had been reported, but the real figure, which included scores of small trading schooners, was many times greater.[1]

The mounting public outcry in Britain and the Australian colonies over the "cannibalistic massacre" of those onboard the *Charles Eaton* gave the new mission particular urgency. In preparation for it, the officers of the *Fly*, which was captained by the experienced Scotsman Francis Blackwood, were made to read a series of gruesome tracts outlining "the treacherous conduct of the natives of the small islands in the Torres Strait." The Admiralty evidently intended this exercise to sharpen the vigilance of the voyagers within this dangerous region; that it might also prejudice the officers in advance of the expedition apparently didn't matter.[2]

One man, however, was determined to resist stereotyped presumptions about the Barrier Reef and its people—the ship's thirty-one-year-old naturalist, Joseph Beete Jukes. Born into an austere but fair-minded Nonconformist family from Birmingham, Jukes had gone to Cambridge to study for a clerical career but been seduced by the extracurricular fascinations of geology. Giving up his clerical ambitions, he worked for a time as a traveling lecturer at Mechanics' Institutes in the industrial Midlands, where teaching hardnosed workers turned him into a religious doubter and a political radical. All his life he longed for "a democratic party that . . . shall come in and sweep away all the relics and dregs of feudalism . . . , reduce the army and navy to a skeleton, remodel the law, the Church, and the whole system of government, abolish all but direct taxation, and . . . commence a new era in the world's history."[3]

A two-year stint in 1839–40 as a geological surveyor in Newfoundland gave him a love of the sea and a respect for the courage and resourcefulness of the Native Americans. If anything, when Jukes left England on the *Fly* for the Barrier Reef in April 1842, he had a marked prejudice in favor of indigenous peoples. "I have always joined in reprobating the causeless injuries sometimes inflicted by civilized, or quasi-civilized man, upon the wild tribes of savage life; and many atrocities have doubtless been committed in mere wantonness, and from brutality or indifference. I have always looked, too, with a favourable eye on what are called savages, and held a kind of preconceived sentimental affection for them, that I believe is not uncommon."[4]

Actually, such affectionate feelings were much less common than Jukes realized. On May 13, 1843, soon after arriving in Australian waters, Jukes was exploring the estuary and hinterland of Wickham's River at Cape Upstart, north of the Whitsunday Islands, with a couple of seamen and the ship's artist. The latter, Harden Sidney Melville, went by the nickname "Griffin" and had become Jukes's close shipboard friend. The group rowed a mile or so up the shoaling river channel, taking potshots at ducks and curlews, when around a dozen native men and women suddenly materialized on the north bank and began trotting toward them along the sandy river plain. Melville later wrote (using the third person for himself):

[T]he men came on, and Griffin will never forget that group of savages
as they advanced . . . To compare a man to a mad dog seems odd . . . but
that foremost savage, that black, brawny, knotty-limbed man-machine,
running like an emu, and flinging up the sand with his indiarubber-like
toes, foaming at the mouth, and howling, much resembled what was
Griffin's idea of a mad dog. His upper jaw was denuded of the two front
incisors; his nose was transfixed by a kangaroo's thigh-bone; the shaggy
hair of his shock head was tied up into a knot with twisted native cord;
his rugged limbs were covered with raised cicatrices; and his body was
besmeared and begreased with filthy pigments. He shone in the sun like
a piece of bright metal, and his feet came down with a heavy thud on the
sand. As he ran he belched out "Ugh, ugh, ugh!" and then uttered a
howl. His adornments were a fillet of grass across what claimed to be
called a forehead (it was an eye-case, not a brain-case), a red smear of
ochre inclosed his eyes and crossed his nose . . . He had a belt around his
waist, a bunch of white cockatoo-feathers stuck in his hair, and carried
an ugly "waddy" in his fist . . . Murderous-looking were all of them, with
fierce, bloodshot eyes rolling wildly.

    With the foam literally falling from their mouths on they came,
leaped into the boat like gibbering apes, and commenced overhauling
the sail, and the oars, and even the persons of the crew, uttering their
outlandish jabber . . .[5]

Griffin and the sailors raised their muskets to fire, but to the young
artist's chagrin Jukes defused the situation with an uninhibited perfor-
mance of clowning and dancing. Soon the warriors were laughing and
imitating his buffoonery. Griffin was unimpressed: "it was noticeable,"
he commented darkly, "that between the acts . . . the rogues showed the
cruel instincts of the savage mind, to be developed so soon as the white
man's weakness had been discovered."[6]

Where Griffin had seen foaming savages, Jukes described "tall, athletic men, bold and confident in their manners, with energetic gestures
and loud voices." In fact, he said, the armed warriors, whom they'd actually met a few days earlier, could hardly have been more genial: "We
saluted our old friends by dancing, on which they began dancing, laughing, and singing, the others sitting still and looking on. As soon as we
had dined we went ashore again, and our friends rushed down to meet
us. Thomas [a young man who'd taken a shine to Jukes] came up, and

Interview with Natives at Wickhams River by Harden Sidney
Melville. Published in *Narrative of the Surveying Voyage of
H.M.S. Fly* by J. B. Jukes, 1847 (National Library of Australia)

embraced me several times, making a purring noise; and whenever a new
face came up, he put his arm round me again, and spoke to him; intro-
ducing me, I suppose, as his particular friend. Ince, Melville, and I went
with them along a path-way down the river, and both tribes followed us.
They were very gentle in their manners and careful of us . . ."[7]

True, a subsequent incident, just over a month later, did test the limits of Jukes's tolerance. On June 25, the *Fly* anchored off Night Island near
Cape Direction, south of the present-day Lockhart River. Some sailors
from their sister survey ship, the schooner HMS *Bramble*, were filing
down a nearby hill, having taken magnetic observations from the summit, when Jukes suddenly noticed a warrior creeping up behind the cox-
swain, his throwing stick hoisted ready to hurl a spear. Jukes pulled the
trigger of his borrowed gun, but it misfired twice and he failed to avert
the tragedy.

According to Griffin:

> . . . poor Baily's shriek of agony awakened the whole party to a knowl-
> edge of the murderous act. The savage, drawing himself up to his full

height, exulted for a moment over the fiendish deed. The geologist, having recapped the piece, again pulled the trigger, and the charge exploded; but it was too late: a fatal destiny had demanded the sad sacrifice.
The murderer escaped . . . Poor Baily, hapless victim of a barbarian's
spleen! It was a cruel fate. They broke the spear short off, for it could not
be extracted. The barbed head had imbedded itself in one of the processes of the dorsal vertebra . . . The poor fellow lingered for a day and
then died. Luckily the vessels lay too far off the coast to afford an opportunity of summary vengeance being taken on the kinsmen or countrymen of the murderer . . . [8]

Jukes himself admitted to feeling an outburst of "mixed rage and grief"
against the perpetrator, as well as a suppressed impulse to exact vengeance
on the whole tribe. He'd never before seen a death inflicted "in any kind
of strife," and he was shaken to the core. He was baffled, too, by the
seeming irrationality of the act. Like Cook before him, and so many
European travelers after, Jukes failed to associate this sudden display of
Aboriginal hostility with the fact that the *Fly*'s sailors had just caught a
swag of fish off the beaches of the clan's estate without asking permission. Jukes's anger over the incident lasted "many days or weeks after,"
but it did eventually fade as he realized that all Aborigines could not be
blamed for the act of one "cowardly" villain.[9]

Joseph Jukes, whom Griffin nicknamed "the geologist," was officially
charged with investigating the geological character of the Great Barrier
Reef and the structure, origins, and behavior of reef-growing corals—
the first scientist ever to be specifically assigned such a task.

Naturally the Admiralty's concern was more practical than scholarly.
By the 1840s it was widely recognized that corals were not inert rocks
but living organisms, although little was known about the cause, extent,
and speed of their development. It was thought that dangerous new reefs
might suddenly appear in places where previous surveys had shown
nothing. The Admiralty hydrographer Francis Beaufort urged the *Fly*'s
Captain Blackwood to remember that he would be dealing with submarine obstacles "which lurk and even *grow*."[10]

It was also expected that a geologist would offer expert advice on suitable sites for future harbors and settlements, and when Captain Black-

wood gave Jukes responsibility for producing the official journal of the voyage, the geologist stressed that he would approach the task as a down-to-earth scientist, conveying "plain fact" and "simplicity and fidelity." He claimed he would eschew any selecting "for effect," or "heightened recollections," or "brilliancy, elegance, or graces of style."[11]

Still, in early January 1843, even this man of plain fact admitted being disappointed on his first inspection of living coral reefs. The fringing reefs off the coral cay Heron Island "looked simply like a half drowned mass of dirty brown sandstone, on which a few stunted corals had taken root." Yet as soon as he broke open some coral boulders that had detached themselves from the main reef and saw their calcareous inner structure, his interest was fired. Jukes decided to throw all his powers of observation and inference into unlocking the mysteries of corals.

The first and most obvious question he needed to answer was how the calcareous fragments of sand, shells, and corals had become "hardened into solid stone," with a regular bedding and a jointed structure like the blocks making up a rough wall. After considering a variety of hypotheses he tentatively concluded that the core structure of these blocks must have been produced inside a mass of loose sand and corals, and that the latter's calcium skeletons had dissolved to make a liquid limestone binding agent. Having then been pounded by waves, the loose exterior of the blocks must have washed away, leaving the solid inner rock exposed.[12]

On undertaking a minute examination of a smaller coral block raised from underwater on a fishhook, Jukes made another important discovery about the character of this strange organic rock—the property that we would today call biodiversity. The surface was studded with a mosaic of tropical coral types: "brown, crimson, and yellow *nulliporae*, many small *actiniae*, and soft branching *corallines*, sheets of *flustra* and *eschara*, and delicate *reteporae*, looking like beautiful lacework carved in ivory." Interspersed with these were numerous species of small sponges, seaweeds, feather stars, brittle stars, and flat, round corals that he'd not seen before.

Breaking open the block, he found, honeycombed inside, several species of boring shells, bristle worms in tubes that ran in all directions, two or three species of tiny transparent marine worms twisted in the block's recesses, and three small species of crab. This single chunk of limestone rock was, he concluded, "a perfect museum in itself." For the first time he allowed a note of excited wonder to creep into his observations, as he reflected on "what an inconceivable amount of animal life must be here scattered over the bottom of the sea, to say nothing of moving through its waters, and this through spaces of hundreds of miles. Every corner and crevice, every point occupied by living beings, which, as they become more minute, increase in tenfold abundance."[13]

Jukes summarized his conclusions for the benefit of Admiralty planners and fellow naturalists, estimating that the Great Barrier Reef extended, with relatively few internal breaks, from Sandy Cape in the south for some eleven hundred miles north, to the coast of New Guinea. It was made up mainly of individual coral reefs, lying side by side in a linear form and running roughly parallel with the coastline, though at distances that varied between ten and several hundred miles. The reefs of "the true" Barrier rose on the outer side in a sheer wall from the great depths of the ocean floor, while on the inner side lay a shallow lagoon scooped out of the coral that had grown up on a subsided landform. The outer reef sections were usually between three and ten miles long and around one hundred yards to a mile wide. They took the form of jagged submarine mounds made up of corals and shells compacted into a soft, spongy limestone rock; this was flat and exposed near the lagoon wall's low-water mark, and higher at the windward edge where the surf broke fiercely and the reef plunged down to the ocean floor.

The sheltered lee side of the outer Barrier, where breaks—and thus passages for ships—were likely to be widest, was generally covered in living corals, but these corals could only survive to a depth of twenty or thirty fathoms because of their need for light. Jukes concluded that a coral reef was actually "a mass of brute matter, living only at its outer surface, and chiefly on its lateral slopes." Alongside these linked linear reefs were a few detached reefs lying just outside the Barrier, as well as a further scattering of inner reefs between the Barrier and the shore. However, at its southern beginnings near Sandy Cape and at its northern edge in the Torres Strait, the lines of reefs were not "true barriers" rising up from deep water, but encrustations of corals growing on shoals or underwater banks and ridges.[14]

For the first time among European commentators, the term "barrier" also carried some positive connotations. Unlike Cook and Flinders and others, who'd seen the Reef solely as a terrible obstacle to navigation— something that prevented access to the shore or escape to the open sea—

Jukes thought of it as a "bastion" that provided the Australian mainland and offshore islands with a protective shield against the massive forces of the ocean. Implicitly he was thinking about the Reef from the perspective of those who lived permanently on its coast, and who benefited from its protection. If laid out dry, he wrote, the Great Barrier Reef would resemble "a gigantic and irregular fortification, a steep glacis crowned with a broken parapet wall, and carried from one rising ground to another. The tower-like bastions, of projecting and detached reefs, would increase this resemblance."[15]

But how did the Great Barrier Reef accord with Charles Darwin's recently published and compelling explanation of how coral reefs came into being? Darwin argued that because reef-growing corals needed light, they could only live in relatively shallow waters. And the only way they could have produced such vast, submerged structures on the bottom of the ocean bed would be by living corals keeping pace with a slowly sinking ocean floor. Those corals enveloped in the deep, dark water would eventually die, leaving behind a mountainous pile of dead limestone rubble. New living corals still close enough to the surface to receive light would grow in a thin crust on top of this.

Darwin's theory seemed at first glance to be incompatible with a local phenomenon that puzzled Jukes. All along the coastline opposite the Barrier Reef he noticed strips of flattened coral conglomerate and pumice stones that were situated behind the beaches, usually some ten feet or so above the highest possible tidemarks. This suggested to him that for a long period of time—say, two or three thousand years—the Australian coast had not undergone any subsidence; it must have remained virtually stationary, with occasional slight movements of elevation.

Yet far from thinking that this invalidated Darwin's theory, Jukes had no doubt that a major subsidence had originally created the Great Barrier Reef, but at a much earlier time than these local elevations. In fact, everything he saw of the Reef and its lagoon convinced him that only Darwin's theory could explain their peculiar topographical assortment of deep and shallow coral structures, shoals, channels, and islands.

He pointed out that the great sweeping curtain of the outer Barrier faithfully followed the curves and flexures of the existing northeastern coastline. To illustrate his point, Jukes offered his readers a compelling hypothetical example. Imagine, he wrote, that we cleared all the existing Barrier Reef corals and raised the intermediate land between the former

Reef and the present coast to a height of around one hundred fathoms, so that this newly raised land emerged just within the line of the present Barrier. If we then allowed reef-growing corals to begin their work in the shallow coastal waters on the fringes of this land, and we then subjected the ocean floor underneath it to a gradual subsidence over a long period of time, we would have the present Barrier Reef.

The fair-minded *Fly* geologist claimed to have ransacked his mind for any alternative hypothesis, but he'd found none that would work. He could only conclude that Darwin's idea "rises beyond a mere hypothesis into the true theory of coral reefs."[16]

For Jukes's own part, his scientific analysis of the geology of the Reef would remain unsurpassed in clarity, brilliance, and originality until the early twentieth century.

Had Joseph Beete Jukes been simply the factual scientist he claimed to be, he might not have appealed so strongly to the mercurial Griffin. Despite Jukes's commitment to plain geological reason, more romantic impulses also jostled within his makeup. In a letter written from Hobart in November 1843, he told his brother-in-law that he'd become a geologist essentially in order "to wander at my own wild will . . . [and] to sigh for the freedom of the open seas or trackless woods of a wild country." So much did he enjoy sailing on the *Fly* that he several times regretted in his letters home that he hadn't gone to sea as a boy to make a career in the navy. Perhaps he saw in Melville the enthusiastic sailor he might have been, for unlike most seagoing naturalists, Jukes loved the life of action and adventure that voyaging offered. Time and again he volunteered to lead shore parties to explore new habitats or to defuse potential skirmishes.[17]

One such near-disaster, on February 15, 1845, arose from Griffin's reckless violation of the captain's orders never to plunder Aboriginal sacred relics. On entering an empty hut at Evans Bay, the northernmost bay on Cape York, the young artist couldn't resist stealing a beautiful emu-feather ornament lying beside mortuary remains, which took the form of wrapped bundles of human bones. It was an action that immediately brought thirty armed warriors rushing to within fifty yards of the hut, "accusing Griffin of desecration of their hearths and homes." By leveling his musket at the leader, Griffin managed to scuttle back to the ship, where he had to apologize to an angry Captain Blackwood for creating a diplomatic crisis.

Jukes soon redeemed his friend's disgrace with an act of considerable bravery. Leading a water-seeking party to the same beach, he was confronted by a large group of irate warriors. "Upon observing this," Griffin recorded, "the geologist, laying down his gun, advanced with open hands, having previously arranged that, should an attack be made upon him, he should throw himself on his face, so that the supporting party might fire upon his assailants. This system of negotiation was successful, and peaceful relations were established." Within a few hours Griffin was able to sketch portraits of the now "friendly and communicative" warriors.[18]

The athletic geologist, tools in hand, cut a dashing figure and he knew how to play to it. He asked his sister in a letter to imagine the *Fly* and the *Bramble* meandering through the Reef's blue waters, and "in the stern-sheets . . . you may put me, with a white shooting jacket, panama hat, luxuriant beard and mustache, gun, and collecting-basket." Griffin soon came to view his older friend as a figure of high romance, like the famous German naturalist-adventurer Alexander von Humboldt. Adjacent to a small drawing of Jukes at work in the coral shallows of the Percy Islands, Griffin penned a lyrical description:

> At low water the reefs about the island afforded a grand field for research to the geologist, who luxuriated amongst them knee-deep and hammer in hand. There were fields of divers kinds of marine vegetation of many colours, some having the appearance of a plateau of variegated penwipers with fancy fringes; corals and corallines branching out into noble terraces, with tints of rose and violet-blue; acres of orange-red brain-stone glittering in the sun, and lying in beds of green sea moss; clamp shells (the *Calme giga*) gaping open with the fish spread out like a velvet cushion, and spotted like a leopard; starfish blue and starfish grey, lying in sandy beds; dogfish and tiger sharks darting about in the channels; crabs and crawfish; and millions of things of life, seen and unseen.[19]

Griffin actually adapted this passage from a long rhapsody to Reef corals and biota in Jukes's journal, which was published in 1847 as the *Narrative of the Surveying Voyage of H.M.S. Fly*, and which the young painter confessed to having open beside him as he wrote his own book. Jukes was certainly worth plagiarizing. He'd concluded his own description of the same underwater "garden" with a fine example of his supple prose: "All these [corals], seen through the clear crystal water, the ripple of which gave motion and quick play of light and shadow to the whole, formed a scene of the rarest beauty, and left nothing to be desired by the eye, either in elegance of form, or brilliancy and harmony of colouring."[20]

As this suggests, Jukes's romanticism ran deeper than the sporting of Panama hats and shaggy beards—in fact, it gradually began to inform his views of the Reef's seascapes and landscapes. When younger, he'd felt a "passion" for poetry, both as reader and writer, being especially fond of Shelley's natural imagery even after he'd outgrown the poet's fuzzy mysticism. Jukes's letters home showed a keen appreciation of romantic aesthetics, and particularly of the fashionable landscape-art theories of the picturesque and sublime. This is not to say that romantic ideas inflected his geological theory, but he admitted privately, if not in his official Admiralty journal, that he also possessed "a poetic temperament." He admired Shakespeare above all other writers, and he favored what he called a "Saxon style" of writing: "nervous, strong, picturesque, and expressive."[21]

Using this "picturesque" style, Joseph Beete Jukes became the first explorer-writer to try to persuade his readers that the Great Barrier Reef possessed a distinctive type of beauty and sublimity. He penned moments of rapture that transcended the everyday hardships of being in Reef country—the baking heat, ferocious green ants, swarming three-inch cockroaches, incessant sandflies, and fever-bringing mosquitoes. "[F]or all these discomforts . . ." he wrote to his sister, "how glorious is a tropical night, or a morning before sunrise!—a cool clear sky, with a gentle breeze fanning your temples, and a delicious dew falling around you; the stars sparkling like gems through the liquid air, and the moonbeams glancing and flickering on the rippling water; and this not occasionally only, but night after night for months together."[22]

During much of their voyage, Griffin, in contrast, grumbled relentlessly about the "intense oppressive heat"; the monotonous "russet-brown and semi-baked foliage"; the dry, barren hills, and tedious swampy mangroves. Yet when reading over Jukes's *Narrative* several decades later, in preparation for writing his own, extremely retrospective account, Griffin was clearly so beset by nostalgia that he included several of his friend's finest epiphanies.[23]

On June 5, 1843, for example, Jukes had climbed a steep hill on Lizard Island, following in the footsteps of his hero Captain Cook, and empathizing with him when he'd stood on the summit "to cast a look on the dangers that yet surrounded him." Griffin, though, was moved more by his friend's description of the ravishing view he'd seen on waking up next morning, when the island was covered in cool mist:

> As the sun rose, the morning mists began to creep up the sides of the hill, at first in light curls, but shortly after in dense folds of vapor, that gathering and sweeping round the summit of the hill, opening and closing here and there, greatly enhanced the beauty of the view, both of our own island and the neighbouring rocky islets, but effectually hindered all surveying operations. Soon after the sun rose, and while his beams were nearly horizontal, we observed a very curious and interesting phenomenon. Whenever a bank of mist rested on the western brow of the hill, and the eastern one was clear, we could see our own shadows on the mist, surrounded as to the head and shoulders by a faint iris or rainbow. By watching attentively, all our movements could be discerned in these spectral figures. On extending the arm, I found its shadow reached beyond the halo that surrounded the head. By getting on a rock, the whole figure was perceptible, and each person thus saw his shadow standing in the air, apparently at a distance of about fifty yards from him, with its head surrounded by a halo of glory.[24]

Though an ardent romantic himself, Griffin didn't realize that his friend had witnessed a Southern Hemisphere version of an eerie optical effect known among European artists and poets as "the Specter of the Brocken." Similar natural projections caused by the sun playing on mist were often seen from the summits of the Brocken, in the Harz Mountains of Germany. By the early nineteenth century the phenomenon had mutated into a favorite romantic metaphor for describing the mysterious operations of the creative imagination. Goethe had used the Brocken as the setting for his "night of the witches" in *Faust*, and the Brocken specter had moved Coleridge to produce his famous definition of the sublime as something so awe-inspiring that it negated all powers of comparison. For the first time ever the specter of the Reef had supplanted the beauties of the Brocken.[25]

Jukes also evoked the sublime during another of his visionary moments on the Reef, this time when standing on its outer edge near Raine Island, just south of the Torres Strait. Gazing down from the hull of a

recently wrecked merchant vessel, the *Martha Ridgway*, he was entranced
by the "the unbroken roar of the surf, with its regular pulsation of thunder." The wild July night on the decks of the wounded ship evidently
reminded him of Edmund Burke's famous definition of the sublime as a
feeling that "operates in a manner analogous to terror."[26]

> A bright fire was blazing cheerfully in the galley forward, lighting up
> the spectral-looking foremast with its bleached and broken rigging, and
> the fragments of spars lying about it. A few of our men were crouched
> in their flannel-jackets under the weather bulwarks, as a protection from
> the spray which every now and then flew over us. The wind was blowing
> strongly, drifting a few dark clouds occasionally over the starlit sky, and
> howling round the wreck with a shrill tone that made itself heard above
> the dull, continuous roar of the surf. Just ahead of us was the broad
> white band of foam which stretched away on either hand into the dark
> horizon. Now and then some higher wave than usual would burst
> against the bows of the wreck, shaking all her timbers, sending a spurt
> of spray over the forecastle, and, travelling along her sides, would lash the
> rudder backwards and forwards with a slow creaking groan, as if the old
> ship complained of the protracted agony she endured.[27]

Jukes grasped Burke's idea that the psychological thrill felt on such
occasions depended on seeing a terrifying scene from a position of at
least provisional safety—though the geologist couldn't help wondering
at the same time whether the *Fly* might not similarly "leave her ribs and
trucks on some such wild reef." In a sense he was feeling both real terror
and its pleasurable aesthetic simulacrum. Still, we can imagine that
Jukes's hero James Cook would have been baffled at how the visceral fear
felt by his crew at the sight of these "sullen reefs" could, some seventy years
later, have become a form of excitement at nature's "grandeur and display of power and beauty." As a coral scientist, Jukes felt a further thrill,
too, from his sense that "the reef . . . on which we stood, was one of
nature's mysteries, its origins equally wonderful and obscure, its extent
so vast, and its accompaniments so simple, so grand, and appropriate."[28]

At the same time Jukes did not shut his imperial eye, making sure to
report as requested on several places that seemed suitable for future

settlement. Here again, his meshing of rational expertise and romantic enthusiasm made him a forceful advocate of the Reef's potential. He thought that the coastland, islands, and waters around the Broad Sound and the Whitsunday Passage had "natural advantages" superior to any other region he'd seen on the Reef, with fertile soils, lush grasses, abundant fresh water, plentiful bays and inlets, and exceptional tidal movements. The latter, he pointed out, meant that small vessels could lie securely in the mud and then be refloated at high tide. Good timber was available for building dockyards, a perpetual sea breeze ensured comfort and good health, and there were "numerous small islands, lofty, rocky, and picturesque in character, and covered with grass and pines, with many small coves and anchorages." These last were also well protected from ocean swells and storms by the Barrier's coral ramparts. That the Whitsunday region is today a mecca for cruising boats and tourists testifies to the geologist's prescience.[29]

He was impressed most, however, when the *Fly* reached the main headquarters of their survey, a subregion at the tip of Cape York where the coast of Australia intersects with the islands, reefs, and waters of the Torres Strait. Evans Bay possesses similar climatic and environmental attractions as the Whitsundays: fertile elevated coastland, abundant timber and water, regular rainfall, cooling breezes, and anchorages "defended . . . by coral reefs." These advantages alone, Jukes argued, would make it a much preferable imperial post to the only settlement existing in the far north at that time, Port Essington. The latter was located in the Gulf of Carpentaria and was fetid, ant-infested, fever-ridden, and too remote for passing ships.

The tip of Cape York, in contrast, offered a perfect refuge for ships passing through Endeavour Strait, as well as a strategic and navigational vantage point from which Britain could oversee "the entire trade of the South Pacific, with the whole of the Indian Ocean." Thanks to the survey by the *Bramble* in February–April 1845, a safe direct passage through Endeavour Strait would take vessels within a mile of Cape York, where a new post could provide easy communications and a coaling station for steamships. Like the Straits of Malacca, Jukes predicted, this could eventually become "one of the great highways of the world."[30]

The scientist in him was also elated to discover that this region constituted a major faunal boundary line: "it was evident that in crossing Torres Strait we were passing from the Australian centre of life . . . into that of the Indian Archipelago." The differences between the two regions could be observed in the marine species, which met from opposite oceanic directions. Torres Strait shells, echinoderms, and reef burrowers, for example, were "generally more brilliant in form and colour, than those on the Australian coast." The much damper climate of New Guinea also prevailed in the Torres Strait, contributing to the creation of rich black soil, "dank woods and jungles," and a variety of cultivable edible species, including coconut palms, plantains, yams, taro, and sweet potatoes.[31]

The mixture of Aboriginal and Torres Strait Islander clans at Evans Bay also suggested that the region served as a boundary line for two different racial groups. The Torres Strait Islanders were markedly stockier and lighter-skinned than mainland Aborigines. They belonged, Jukes thought, "to the great Papuan race, which extends from Timor and the adjacent islands through New Guinea, New Ireland, and New Caledonia, to the Feejee Islands."[32]

The Fly's first meetings with Islander peoples—from the islands of Mer, Erroob, Oomaga, Dammood, and Masseed—had been conducted under a cloud of suspicion. These were, after all, the cannibalistic "tribes" denounced in the Admiralty tracts about the Charles Eaton massacre. Jukes, like most of the officers, had read Captain Lewis's account of rescuing the two boys, Ireland and D'Oyley. It was a story that affected him personally, because a Wolverhampton family friend on board the ship had supposedly been "eaten by the natives." Griffin, as usual, could be counted on to exercise supererogatory vigilance. He nicknamed the two leading Erroob traders "Mr. Murderem" and "Clubbrains," and speculated darkly that it was only the Fly's guns that "held in check . . . more than one savage, who, moody and morose, brooded gloomily, with lowering look and murderous scowl, on what he would do with that cruel stone club if he dared."[33]

By contrast, it took Jukes only hours to warm to the young men, women, and children who everywhere walked up "unarmed, [and] with the utmost confidence," holding coconuts and turtle shell, waving conciliatory green branches and shouting "poud, poud, poud . . . meaning Peace! peace" as they advanced. Every shore visit added to his confidence. Crowds mobbed him excitedly, yelling out his nickname "Dookie," and on Erroob Island he formally exchanged names with "a fine straight-limbed and graceful young fellow, called Doodegab," whose chaste, petticoat-wearing sisters were persuaded to stroll with the geologist arm in arm. Jukes was impressed, too, by senior Islander men like Old Seewai, Mammoos, and Old Duppa, whose tribal influence seemed to derive from a combination of wealth, generosity, courage, and wisdom. Old Duppa, who'd rescued the two shipwrecked *Charles Eaton* boys in exchange for a bunch of bananas, even managed to win over Griffin with his "benevolence" and his "affectionate" inquiries about the well-being of their two adoptive sons, Wak and Uass.[34]

Everything Jukes observed seemed to contradict the Islanders' reputation for savagery: the elegant villages studded with fine houses, groves, and temples; the cultivated gardens of yams and plantains; the magnificent twin-hulled sailing canoes; the elaborate collections of crafts, fabrics, carvings, weapons, and trade goods; and the mellifluous languages the Islanders were keen to teach the visitors.[35]

Jukes the naturalist was fascinated by the Islanders' sophisticated taxonomies of local biota. When he asked Mammoos the Erroob terms for a variety of shells, the old man cited "almost as many names as there were genera, and for some species of one genus he had different names." The taxonomic principles used for each category of shell were, Jukes discerned, based on "general form," such as those having wrinkled mouths or longish shells. As a result, the naturalist concluded, "there were many more distinct names for the different shells than we have in common English."[36]

Only once did the *Fly* sailors experience an episode of hostility in the Torres Strait. Returning, during the same voyage, to Erroob in May 1845, they were startled to be greeted with a volley of arrows fired by individuals who'd earlier treated them as honored guests. Sensible Jukes soon discovered that the two most influential men on the island, Mammoos and Seewai, had quarreled during their absence and placed various island sites under taboo, or "galla." One of these sites happened to be Beeka, the *Fly* crew's usual beaching site. As a result, Jukes explained, "our landing there was an offence against their customs, for which the arrows were shot yesterday." After mutual explanations and apologies, both warring groups pledged to restore their former good relations with the visitors. An island-wide peace followed, which included a reconciliation between Seewai and Mammoos.[37]

Shortly after this, even the suspicious Griffin succumbed to the beauties of the island and its people. An afternoon walk to Seewai's village became for him "a romantic adventure" in "a fairy glen," as "the sky, of intense blue, inclosed a leafy tracery of tropical richness, hung with luminous loveliness over an enchanting scene." Inspired by this loveliness, one of the *Fly* sailors began to sing a traditional folk tune—"Through the wood, through the wood, follow and find me"—to which Old Seewai "listened with rapt attention and delight." Eventually, Griffin recalled, "the good old native" led the way back to the beach.[38]

A month later, in a letter written from Erroob to his sister, Jukes summed up the prevailing attitude of the crew toward the Islanders: "Altogether we have been greatly interested and amused with these people, and like them much; and were it not that they have an unfortunate predilection for collecting skulls, no fault could be found with them." He even offered a rational explanation for that gruesome habit, which "proceeds," he explained to his sister, "from an idea that of all the skulls they can collect during life-time the owners shall be their servants and followers in the next world. Of course to have a white man as a slave would be a great honour."[39]

In the same letter, Jukes admitted that the likelihood of receiving orders to return home any moment was weighing on his spirits. Their long sojourn in the Great Barrier Reef had eroded all his sense of belonging to England, which now seemed like "a foreign country." "Were it not for my mother and yourself and my other relations," he confessed, "I should feel little wish to see it again, so strange and remote does it now seem to me."[40]

By the time he prepared to depart from Sydney in early October 1845, his feeling of impending loss had deepened into full melancholy. The idea of resuming a "quiet domestic existence" loomed like a sentence of slow death. Better a native spear in the back any day. "Heaven send I may never have to die in my bed by decay or disease, but may meet the 'grim feature' face to face in the free air, and with my blood warm." Griffin, who'd spent their refitting time in Sydney undertaking a series of rollicking up-country adventures with wild colonials, echoed these sentiments exactly.[41]

Three years later, Jukes was given the opportunity to return to Australia as a geological surveyor, but his new wife was reluctant, fearing the reputation of the remote continent. Instead he took a job working on a geological survey in Ireland, where it was said that the harsh weather, lonely conditions, and the effects of being thrown from his horse caused a gradual breakdown of his body and mind. He died in 1869, only two years after his old friend Griffin published a sparkling tribute to their adventures together as daring explorers of the beautiful Barrier Reef.

Joseph Beete Jukes, "the geologist," was a pioneer of three major traditions of Barrier Reef thought. He was the first scientific analyst of the Reef's geological origins and coral structures, the first professional-style ethnographer of Indigenous Reef cultures, and the first European writer to appreciate the Reef's distinctive romantic beauties.

As an ethnographer, Jukes anticipated the fair-minded Oswald Brierly, whom we will encounter in the next chapter. As an observer and writer, he greatly extended the aesthetic observations of his predecessors Cook and Flinders, and it is surely Jukes's classic version of the sublime as a kind of thrilling awe that today's Reef divers, filmmakers, and sailors still try to convey, even if they're not always aware of its philosophical origins. But it was primarily as a scientist that Jukes influenced his successors. And although Charles Darwin never saw the Great Barrier Reef himself, it is no exaggeration to say that modest Joseph Beete Jukes did the geological work there that Darwin would have liked to, and which even he could not have bettered.

PART TWO

Nurture

5

HEARTH

Barbara Thompson, the Ghost Maiden

SOME FOUR YEARS AFTER JOSEPH JUKES LEFT Australia, a young woman living on the southwestern Torres Strait island of Muralag (Prince of Wales Island) heard news that made her heart sing. Giom, as she was known, learned on October 13, 1849 that a large ship had anchored off the nearby mainland beach of Podaga (Evans Bay). No ordinary vessel, it was that rare phenomenon a *marki angool*, or ghost ship, filled with the wraith-like white creatures that Aborigines and Islanders turned into when they came back to earth after death.

Giom was herself a *marki naroka*, a ghost maiden. Five years earlier she had been rescued from a shipwreck by three young men, whereupon an elder of the Kaurareg people and his wife recognized by her chin and eyes that she was their daughter Giom, who'd died by drowning and had now returned to them from the sea. The elder, Peaqui, and his wife, Gameena, adopted the young white woman.[1]

On this October day of 1849 one of her rescuers—kindly, ugly Thomagugu, a Gudang Aboriginal man from the mainland—came to the island to spread the news of this remarkable ship. The Gudang and Kaurareg intermixed somewhat, and Thomagugu had relatives as well as friends on Muralag. As his adoptive sister, Giom was one of the first to be told the news, and she also overheard him urging the Kaurareg to send an expedition to the ship for gifts of knives, axes, cloth, and bottles.

Knowing that some of the Kaurareg, including another of her rescuers, Boroto, would be opposed to her visiting the ship lest she decided to stay, Giom asked her guardian and sister, a senior woman called Urdzanna with whom she lived, whether she could go. The Kaurareg thought of Giom as a trophy. Her presence brought them status, but being a ghost, she was not completely one of them. Pretending to be ill and in urgent need of medicine from her fellow *marki* on the ship, Giom promised to bring Urdzanna a knife from the sailors. She assured Urdzanna's husband Gunage, a wealthy and influential boat owner, that she had no intention of staying on the ship, and was given permission to go.[2]

Next morning at eight o'clock, the expedition set off in four large sailing canoes. Giom rode with the boat's owner, Old Sallali, his wife Old Aburda, and their three grown-up children—two sons and a daughter. The strongest of their paddlers was Boroto, a burly young warrior who, as well as being one of Giom's rescuers was also Gunage's younger brother. Because there was no breeze, the fleet drifted, and the men scanned the unruffled surface of the water for signs of turtle, which were liable to be copulating at this time of year. Soon a single turtle was sighted and captured, after the excitement of which the fleet headed to a small sandy beach on the mainland where their Gudang friends had lit fires to welcome them in.

As soon as they landed, at about eight o'clock in the evening, the Gudang huddled around, talking excitedly about the knives, *bissikara* (biscuits), and shirts that could be obtained from the *marki*. The Gudang recommended that the Kaurareg base themselves on the adjacent island of Wamalag (York Island), a standard meeting place for traders which had the additional advantage of six good water holes. The island also looked over to Podaga, where the *marki* congregated each day.

That night on Wamalag, Giom couldn't sleep for excitement and nervousness: "I had not eaten anything scarcely for two days for thinking of the vessel, how I should get off." Lying on a soft pile of grass and her woven mat, she listened to the noises of the men talking and joking around the fire. Boroto and a few friends were smoking native tobacco in their long bamboo pipes and intermittently toppling over on the beach in a stupefied heap. Other men chattered incessantly about the prospects of trading with the *marki*, while others again were sleeping in their canoes for fear of being surprised by enemies.[3]

Around 2:00 a.m., a sudden scare swept through the camp when one of the elders was visited by a prophetic dream that they were about to be massacred by their ruthless mainland enemies the "Yegillies" (Gumaku-din clan). Most of the Kaurareg, including Giom, leaped into their canoes and paddled away from shore. Returning when the scare had abated, Giom ate some roasted turtle eggs and at last fell asleep.

Early the next day, after the men had gone off in the canoes to hunt turtle again, a woman friend agreed to accompany Giom up a hill on the eastern side of the island so that she could look out at the ghost ship. Again the castaway found herself quizzed about whether she intended to stay with the *marki*, and she reiterated her need for medicine, adding the reassurance that "I was too black. They would not have me now." Later in the day, when the men returned with a good catch of turtle, they handed it to the women to cook, then went back to their canoes to visit the ship. Giom, ordered by the senior women to remain behind to help with the cooking, was told she might be allowed to go to the ship the next day. Once the canoes were gone Giom began to cry, asking a friend crossly "what they meant at keeping me back from my people."[4]

Clambering up a rock behind the camp, Giom began angrily stripping pandanus leaves with a piece of sharpened hoop iron, recalling a melancholy occasion some years earlier when she'd failed to get a passing ship to stop, even though it was nearby and the sailors on deck could obviously see her. Evidently thinking she was a native woman, they ignored her, and as the ship disappeared Giom lay down by a water hole and wept, for hours, until night fell.[5]

Still, when the Kaurareg men returned that evening, she was heartened that they could talk of nothing but the *marki* ship and the booty they'd obtained. Thomagugu, her thoughtful brother, gave her a gift of a small pipe with a picture of a ship on it that looked like a man-of-war. The rest of the men began to quarrel over which of them should receive the reward for taking Giom to the ship the next day, until Thomagugu silenced them with an angry reminder that he was the one who'd held her up in the water when she was drowning.

Yet even this sharpened interest in obtaining *marki* goods did nothing to precipitate action. The next day passed exactly like the last, except that this time when the men returned from hunting, Old Aburda, the only senior woman who could serve as a chaperone, delayed Giom's visit a further day because she felt too tired for the short voyage across the strait. Giom now pinned all her hopes on Thomagugu and his wife: he'd always said she would one day be allowed to go back to her people, "but most of the men and women said that I should die amongst them."[6]

The following day, October 16, stretched out with the same agonizing slowness. While the men were hunting, Giom busied herself collecting wood, heating stones on a fire, and cutting up turtle for baking. When the canoes returned she bathed in the sea in preparation for a visit to the ship. She had just finished making a shade shelter in the sand for Aburda, when the old lady suddenly walked down to the water's edge and called her to look at some white men who were shooting birds over at Podaga. Seeing this, the Kaurareg men jumped into their canoes and headed for the beach.

Giom begged Aburda and a few other women to go with her, but they were too afraid of the sporadic gunfire, as indeed were many of the men. Soon the Kaurareg canoes were already halfway across the small strait, and only a single Gudang canoe, carrying Thomagugu, remained. Even this vessel had pulled out so far from shore that Giom had to wade out up to her breasts to catch it. Grabbing onto the side, she was pulled along out of her depth until the skipper, Old Den, hauled her aboard: "I came off in such a hurry and was so frightened that I left my basket and large *dadjee* [grass skirt] and everything just as it lay [on the shore]."[7]

After beaching at their usual mainland spot, the Kaurareg men walked straight to Podaga, where some white men were washing clothes at the water holes. Straggling behind with Thomagugu and a few others, Giom came across a group of mainland Aborigines walking along with four white men, but she was so browned and blistered by the sun that they strolled past her without noticing. Summoning her courage, she called out to them in English, "I am a white woman, why do you leave me?"[8]

Abruptly the sailors gathered around. "Thomagugu began to talk to them before I could speak, telling them in his own talk how I had been wrecked and how he had taken me up out of the water. I stopped him and said, '*komi arragi arragi atzir nathya krongipa*'—Friend, hold your tongue, I know what they are saying." Gathering that she was a Scottish girl called Barbara Thompson who'd been shipwrecked and rescued by the natives, the sailors yelled out to another man at the water hole, "Scott, Scott, come here. Here's a Scotch girl."

> [Scott] took hold of my hand and led me along to where the men were washing. He was like a guard to me. I could not understand the other men as I could him. When we reached the washing place, he took me into the bush and with another man . . . washed me and combed my hair, and dressed me in two shirts, one below as a petticoat, the other over my shoulders. I was so ashamed when I got to the washing place that I did not notice what men were there. But this Scott was a friend to all of them. He took hold of me so bravelike. As I went along I could hardly speak for crying.[9]

On that afternoon of October 16, 1849, Oswald Brierly, a thirty-one-year-old artist on the British survey vessel HMS *Rattlesnake*, was practicing shooting with a few shipboard friends on the beach at Evans Bay when an officer suddenly ran up to them, shouting that "the blacks have brought a white woman up to the beach."[10]

Without waiting to hear more, Brierly sprinted to where he could see a mixed group of Aborigines and sailors gathered around a woman aged, he guessed, about twenty. (She was probably eighteen, but looked older.) She'd been given two shirts to wear, having appeared at the water hole naked, "with the exception of a narrow fringe of leaves in front."[11]

That evening Brierly took up the rest of the story in his journal. "She sat on the bank with her head hanging down and had a tin plate with some meat and a knife and fork, which the men had given her on her knees before her. One Black sat close to her with his arm passed behind her, two others were standing close to her. Her manner was very curious and she replied to our questions something in the manner of a person just waking from a deep sleep."[12]

Barbara answered his queries in a halting mixture of Kaurareg and English, drumming her forehead in anguish. "I forgot English since I saw my country. I sing the song I knew when I lay down at night, to remember it." The sailors nevertheless gathered that she was the daughter of a Sydney tinsmith, and that her husband had drowned on a reef when their cutter was shipwrecked.

When quizzed by Brierly whether she wanted to stay with her clan or join the ship, she replied, "I am a Christian," explaining incoherently, "I saw my people today. I sat on the island all day looking at the ship. I said tomorrow *kusta kalloo* and with your people. And I told them [the Kaurareg] lies to make them take me in the vessel. I saw my country people in their little boats—*arawa gool*."[13]

As Barbara was rowed out to the ship on a jolly boat, accompanied by a very insistent Boroto, among others, Brierly picked his way through her garbled mixture of Kaurareg and English to learn the essence of her story. Her family had migrated from Aberdeen to Sydney while Barbara was a child. When still a girl she eloped with a sailor to Queensland, and the two of them, along with some other sailors, had been trying to salvage whale oil from an old wreck in the Torres Strait when they were ship-wrecked themselves. Her husband and the other sailors had drowned, but she'd been rescued by turtle hunters from Muralag, who'd looked after her well. After that she'd been adopted by the Kaurareg, with whom she'd lived for the past five years.

When the jolly boat reached the *Rattlesnake*, Captain Owen Stanley welcomed the young woman warmly, despite his private trepidations about having her on his ship. He fed her apple pie, gave her a cabin segregated from the rest of the sailors, and had the doctor tend her burns and infected eyes. Thomas Huxley and John MacGillivray, the ship's inquisitive young naturalists, who both thought the castaway "not bad looking," managed to question her briefly. But Captain Stanley made it clear that Mrs. Thompson would in future dine only with him and Brierly. The artist would also have the exclusive right to visit her cabin to conduct interviews.[14]

Nobody was surprised at this news: Oswald Brierly hailed from the same gentry circles as the captain, and he'd been invited onboard the *Rattlesnake* as Stanley's personal friend and companion. Brierly himself put the situation more tactfully: "myself having no duty [aboard] the ship, I could divert what I chose to writing down her accounts and employ a larger part of every day to writing down whatever she remembered of her island life and the [customs] of the natives."[15]

There was also a cogent practical reason for allocating Brierly the role of interviewer. During the relatively short period that the artist had been aboard the *Rattlesnake*, he'd built up an exceptional rapport with local native peoples. While his companions' occupations consisted of surveying or natural-history projects, "mine," he wrote, "was to indulge my fancy for talking to the natives and to gain if possible some ideas of their peculiari-

ties of [mien] and appearances, etc, an employment which on board has
received the name of 'niggerizing.' "[16]

Brierly's fascination with Indigenous people and culture went well
beyond the standard Enlightenment vogue for collecting taxonomies and
artifacts. He seemed to admire the Aborigines, and to like many of them
individually as "friends." In the short time since the *Rattlesnake* had landed
at Evans Bay, he'd made a strong impression on the mixed communities
of mainland Gudang Aborigines and visiting Torres Strait Islanders from
Nagi (Mount Ernest) and Muralag. These groups gathered annually at
the neutral ground of Evans Bay to trade food and artifacts and to hunt
for turtle during the last weeks of the bountiful dry season before the
onset of the testing northwest monsoon.

Brierly's amiable personality, long thin face, and distinctive beard
had earned him immediate recognition among the locals as the *marki* of
a dead Kulkalaig man (of Mount Ernest Island) named Atarrka or Tarrka.

Native huts, Evans Bay, Cape York, Novr. 1849. Watercolor. In
Narrative of the Voyage of H.M.S. Rattlesnake: Vol. I
(Mitchell Library, State Library of New South Wales)

It was a name and role that the artist readily adopted, and now his ghostly status gave him a special kinship with the young castaway.

Such was Brierly's commitment to learning the local dialects and getting to know the Reef peoples that he spent hours sitting cross-legged in the sand, gossiping as he collected and memorized vocabularies. Though he was an upper-class Victorian Englishman, nothing about the natives' habits or values seemed to faze him. He was happy to joke, clown, and play tricks with the young men; he strolled unarmed with spear-carrying warriors into the bush; he sweated for hours in the blazing heat to capture his black friends' likenesses on paper; he threw himself into uproarious games with the children; he sampled the most challenging of foods on offer; and he joined energetically in dances, songs, egg hunts, and spear-throwing competitions. He also showed genuine admiration for the men's ability to track and catch the near-invisible molluscs lying in the soft tidal mudflats and shallow water ripples, and he praised the women's intimate knowledge of the whereabouts, among a tangle of brush and stones, of delicious bush fruits and yams.[17]

Brierly became so popular among the clan that he had a brotherlike relationship (*cotaiga*) with three young men, as well as the patronage and friendship of many seniors. The latter included an older woman called Baki, his "self-constituted mother," who boasted a much younger husband and exercised marked authority over all the clan groups at Evans Bay. Brierly thought her a fount of good sense. He nicknamed her "Queen Baki," showered her with gifts, invited her on board the ship to meet the captain, and responded to her enthusiastic embraces with reciprocal affection and good humor.[18]

How Oswald Brierly became so open-minded is unclear: such freedom from racial condescension was rare in Britain and the colonies, and aboard the *Rattlesnake*. Although the ship's two naturalists, Huxley and MacGillivray, showed a keen interest in ethnography, even they had been prejudiced against the Barrier Reef by the negative publicity of Eliza Fraser's "captivity" and the *Charles Eaton* massacre, which had prompted the survey expeditions of both the *Fly* and the *Rattlesnake*. There was nothing, either, about Brierly's background to suggest social unorthodoxy. He was born in Chester in 1817 to an old English upper-middle-class family, and his father, a doctor and amateur artist, had en-

couraged the boy's education at a London art school, where an interest
in ships led him to specialize in maritime art.

An associated passion for sailing had then led him to Australia, and
it was while managing an isolated pastoral and whaling business for five
years at Twofold Bay in New South Wales that he first came to know
Aboriginal people. He sketched their portraits, landscapes, and boats,
and made several friends among the Aboriginal whalers, whose company
helped him survive the solitude.[19]

To make such intimate connections, he'd needed to cross the barriers
of language, something Brierly tried to do wherever he traveled. His
journal shows him going to inordinate trouble to capture the exact linguistic meanings and pronunciations of both Gudang and Kaurareg
dialects, including sometimes pretending deafness so that difficult words
would be repeated slowly and loudly in his ears.[20]

His love of sailing was another factor in his fascination with the native peoples he met at Evans Bay. Brierly's journal is filled with sketches
of the rudders, hulls, rigging, and ornamentation of their giant sailing
canoes. So dazzled was he by the workmanship and "graceful form" of
one such boat, the *Kyee Mareeni*, or *Big Shadow*, owned by a Kaurareg
elder named Manu, that it became his boating ideal: "I had long admired but never till now seen anything that realized so much the idea of
beauty." He sketched the craft over and over, and it was later to feature
in both an oil painting and a wall mural of his. The sight of the Kaurareg
vessels drifting by in the still evening waters of the northern Barrier Reef
touched his deepest aesthetic instincts: "As they drew the canoes in the
water and left the white sandy beach, they were swept slowly downwards
by the current and soon became lost in the dark reflections of the rocks
and hanging foliage of the side of the island. A broken light made by
their paddles in the water . . . at times just indicated their place in the
dark shadow."[21]

Brierly's artistic ambitions are a further clue as to why he pushed so
hard to get beneath the surface of the Aboriginal and Islander peoples he
met. His writings and sketches contain minutely observed details of the
physical appearances, personalities, and idiosyncrasies of friends like his
young *cotaig*, Belidi. He strove to capture subtleties of character—exactly
the opposite impulse of his artist counterpart on the HMS *Fly*, Harden
Melville, who'd thought "savages" suitable only for comic caricature. Such
crass attitudes provoked Brierly to advise would-be painters like his friend

Prince Wales's Island Canoe "Bruwan." In *Sketches on Board*
the H.M.S Rattlesnake. Made during the Coastal Survey of the
Passage between the Great Barrier Reef and the East Coast of
Australia, ca. 1848 by **Oswald W. B. Brierly** (Mitchell Library, State
Library of New South Wales)

Thomas Huxley "to record what actually passes under your observation—
any characteristic traits or circumstances which transpire under your
eyes should be written down while the impression is fresh and as quickly
after their occurrences as opportunity may allow—in doing this you will
be constantly surprised to find the savage so utterly different from what
your preconceived ideas would make him."[22]

For the same reason, Brierly was meticulous about capturing the
exact words, inflections, and meanings of the linguistically confused
Giom-cum-Barbara Thompson. Early on in their shipboard interviews—
when her recall of English was limited and she, an illiterate woman of eighteen, seemed shy of the rather grand figure of the captain's companion—he
prompted her gently with specific questions about the everyday life of

the clan. His method was to transcribe "with as little deviation as possible from her own words and mode of expression. And to ensure its accuracy I . . . read it over carefully to her several times, making any slight alterations in pencil before finally writing it all in." Sometimes he would also return to particular accounts after several weeks, to see whether her recollections or perspective had changed in the interim.[23]

The ultimate credit for the quality of the information supplied must of course go to Barbara herself. Brierly thought her "remarkably observing," and illiteracy probably sharpened her powers of recall. Even cynical Tom Huxley and abrasive Jock MacGillivray praised her intelligence, honesty, and courage. And because her attitude to the Kaurareg combined a broad sympathy with a slightly conflicting wish—as we'll see—to convey some degree of emotional distance, she generally gave Brierly a balanced evaluation of clan life and culture.[24]

Her greatest knowledge, naturally, was of everyday work patterns and the social relations of the women. She recounted their modes of childbirth and nurture, including the vogue for massaging the head of every newborn baby into the beautiful elongated shape of a remora fish, and described their methods of making baskets, mats, grass skirts, and fish traps using the leaves of the indispensable pandanus plant. She listed their modes of treating illness with bush medicines and controlled bleedings, and explained the intricate network of kinship structure and family relationships, outlining taboos and norms as well as the more mundane interactions between husbands, wives, children, and lovers.

The island of Muralag, with its rocky soil, scrubby vegetation, and single coconut tree, made foraging a challenge for the Kaurareg women. Barbara described for Brierly their role as collectors, preparers, and cooks of varying foods in accordance with the two great seasonal shifts of the year. During the dry season (*iboud*), from June to October, when the southeast trade winds prevailed and turtles, fruits, and fish were relatively abundant, the clan would travel to other islands or mainland spots where the men would hunt and the women would search for yams (*coti*), shellfish, and water. Bladders of turtle oil were then mixed with crushed yams to make a much prized and portable mash (*mabouchie*). In each camp the women would dig large ovens, lining them with stones, for cooking turtle and dugong. They then carved and arranged these delicacies in elaborate patterns of distribution according to age, status, and gender, leaving nothing unused.

From December to April, during the great wet of the northwestern monsoon (*kuki*), the clan would gather in regular camps near sources of edible mangrove pods, which the women would prepare in a mash mixed with wild beans called *beu*. At this time, when turtles and dugong were scarce, the Kaurareg experienced a relatively static and claustrophobic period, crouching "like hens" in long narrow huts, enduring weeks of incessant tropical rain and occasional bouts of fever (*doopoo*).[25]

Although she had lived in that world as a mere ghost, Barbara also managed to open up windows into masculine life. She gave Brierly glimpses of the secret rituals of boyhood initiations, and of sorcerer (*mydallager*) curses and magic. She described the different forms of hut building; Kaurareg patterns of trading, diplomacy, and warfare with other Islander and mainland clans; and the cherished skills of canoe building and sailing. She knew the techniques of weapon making and ornamentation, fire clearing, and yam cultivation, and the smart ways of trapping fish in stone weirs and creek nets. She knew, too, the cunning skills needed for finding and killing dugong and turtle. The latter, at which Boroto excelled, included attaching magical potions to canoe prows to entice green turtle (*soolah*) to the surface, and an ingenious practice of using live remora sucker fish (*gapoo*) threaded on thin rope. The fish were tossed back into the water to locate flatback turtles by clamping onto their shells.

Specifically quizzed by Brierly, Barbara described in detail the clan's elaborate death and mortuary rituals. She recounted myths of the Kaurareg's origins, rich foundational stories that blended human and animal elements, and she told the spiritual stories associated with sacred local places and objects. She talked of the modes of singing, dancing, and game playing, citing satirical songs composed about visiting white sailors and their ghost ships.[26]

Some of her revelations must have shocked even the tolerant Brierly, but whether she was telling him of the awful injuries that some husbands inflicted on their wives or the practice of smothering girl babies who'd been conceived before marriage, he recorded her matter-of-fact words without emphasis or comment. Barbara showed no reticence about describing cannibalistic rituals performed on the decapitated and baked heads of tribal enemies. She told the gruesome story of a revenge raid by Kaurareg warriors against the Gumakudin. It had been well deserved, she said: the Kaurareg's much hated and feared mainland rivals had, without any provocation and "in such a sly way," killed and mutilated a senior Kaurareg man—the father of Boroto—on a lonely Muralag beach. In celebration of the successful retaliation, which resulted in the warriors bringing home six fresh heads, "[our warriors] took the eyes and then cut the flesh from round the eyes, the men who cut it passing bits round to the rest. When they eat it they would throw their heads back with their mouths open, holding the bit ready to drop in, [and] call out to their wives, *Areen, idoo eenama*, 'Our food, look at it.'"[27]

Later, as Barbara became more fluent and confident, Brierly encouraged her to relate stories at her own pace and length. Many of her revelations flew in the face of what most Europeans believed they knew about native Reef peoples. Anyone who'd read of Eliza Fraser's torments or the *Charles Eaton* massacre would have expected tales of animality: lack of family affection, indifference to beauty, preoccupation with material goods, and intellectual and moral nihilism. Barbara, by contrast, took for granted that personal relationships within the Kaurareg could be harsh or loving, depending on individual personality and circumstance. She gave examples of the anguished remorse shown by some men who'd mistreated their wives, and of the retributions exacted from them for doing so. She amplified Brierly's impressions of the clan's caring attitude toward children, and told of the solicitous way Old Sallali had looked after his wife Aburda when she was pregnant. Barbara spoke of the doting pride of a young man called Dowoothoo in the abilities and status of his senior wife, Baki—"a pattern," Brierly noted dryly, "to husbands all over the world."[28]

Barbara poured out stories illustrating the love and care showered on her by adoptive relatives and friends. Huxley observed that the Kaurareg had "treated her quite as a pet." Male elders, like her father Peaqui and her uncles Old Manu and Old Sallali, pampered her with the best portions of dugong and turtle, and they protected her health by persuading the womenfolk to exempt her from the heavy work of collecting food. Instead she was given the lighter duty of looking after the camp children while their mothers were working. These same senior men presented her with a yam-stocked garden on the tiny island of Nuripai, which they named after her—an exceptional honor. Brierly recorded a typical scene of avuncular affection when Old Sallali visited Barbara on the *Rattlesnake*: "[He] comes and sits cross-legged by her, talking in such kind tones to the white woman, calling her his child, and looks with quiet wonder as she displays before him all the gowns which she has been making up from cotton handkerchiefs in the piece, the only thing on board which would serve."[29]

The women—Urdzanna, Aburda, Gameena, and Baki—had been just as attentive as the menfolk to Barbara. When she was ill with *doopoo*, they bled her to reduce the pain and fever, and gave her bush medicines. If there was sometimes an affectionate condescension in their kindness, it came from her being "only a *marki* poor thing." On the death of some of her kin, they excused her failure to demonstrate a proper depth of emotion by cutting herself and wailing for hours. Otherwise reprehensible, her behavior was forgiven because *marki* "don't cry, they have no feeling." After all, they were not "real people."[30]

This didn't stop these same women from grieving piteously at Giom's departure. She told Brierly that the women were crying for her at the camp, "as if I was dead." During the nine weeks that she lived onboard the *Rattlesnake* before it set sail from Evans Bay, they brought daily gifts of cooked turtle eggs, yams, and woven baskets; they paddled under her cabin porthole weeping and begging her to return to Muralag. Brierly witnessed one such visit from her wistful mother, Gameena. "[She] showed the greatest joy at seeing Mrs. T. at the port and stood up in the canoe till she might take hold of her hand, which she kissed with great affection, at the same time showing a shell which had belonged to Mrs. T. while on land . . . in which she had bored a hole and now wore round her neck as a remembrance, saying *Giom, ye noosa eena* – 'Giom, this is yours' and at the same time kissing it."[31]

So proud was the whole clan of Giom, in fact, that they'd been prepared on one occasion to risk a war by hiding her from an armada of sixteen Badu war canoes. These had arrived uninvited on Muralag one evening, intent on abducting the young *marki naroka* to their island. The sinister fleet carried two hundred warriors, most of whom had bows, spears, and bamboo beheading knives. The Badu leaders' glib claim that they'd come "for pleasure, on a sort of pleasing party" fooled nobody. They flourished gifts of turtle for Giom, and their leaders asked immediately to be taken to "shake hands" with this "*kwari guri*—strange creature."

A Badu woman named Nuadji, who'd been sent ahead as an envoy, quietly took Giom aside and whispered a secret pitch from the Badu. Unlike "stony Muralag," Nuadji said, the island of Badu was blessed with abundant coconuts and bananas. If Giom moved there, she would be fed lavishly and even given the rare gift of *ogada*, or totemic status, to protect her from any unwelcome advances from men.[32]

Eventually it emerged that what the Badu really wanted was to marry her to their own resident *marki*, a sailor they called Weenie or Gienow. Probably of Malay or Portuguese extraction, he had arrived on Badu after a shipwreck some ten years earlier, whereupon he was adopted and protected, and as a result was said to be "owned" by two brothers. He quickly became indispensable among the larger community for his skill as a repairer of canoes.

When Weenie later visited Muralag without an accompanying war party, he explained to Giom that he was happy living with the Badu and hoped never to leave them. But as a *marki* he was not permitted to marry formally within the clan. Several older widowed women were allowed to live with him, and even to bear his children, but many of the married Badu men remained suspicious of his intentions toward their wives. If he were to marry Giom, such jealousies would be allayed, for, as he gestured to her, "we are like the same people." She warmed to the tall, middle-aged castaway with a pockmarked face and gentle manners, yet she didn't want to leave her Kaurareg people for an unknown and possibly bloodthirsty new clan.

Nuadji's eagerness to lure Giom to Badu Island led her to let slip the visitors' plans, which included a willingness to slaughter Giom's Kaurareg family to secure her transfer. "[Nuadji] said if I would say that I would go, they would bring canoes over to me in spite of the Kauraregs and kill some of them besides. I was very much frightened when she told me this [and] would not go with her." With the help of friends, Giom hid among the rocks and brush on the other side of the island, and despite the strenuous efforts of the Badu to intimidate the clan into giving her up, the flotilla eventually had to leave empty-handed, two nights later.[33]

Crusty Jock MacGillivray, who understood both Scots idiom and the Kaurareg language better than any of the other sailors on the *Rattlesnake*, including Brierly, added in his journal account of the voyage a startling variant to Barbara's story. Her most insistent and aggressive protector was the warrior Boroto, who as one of her rescuers had earned the status of adoptive brother. Barbara called all three of her black rescuers, in her strong Scottish accent, "my boothers," but MacGillivray claimed to have learned through shipboard conversations with Boroto that he was more than this. "One of these blacks, Boroto by name, took possession of the woman [Mrs. Thompson] as his share of the plunder," he recorded in his usual blunt fashion; "she was compelled to live with him, but was well treated by all the men . . ."[34]

Boroto pleaded continually with Barbara to leave the *Rattlesnake* and return home, prompting MacGillivray to elaborate: "Her friend Boroto, the nature of the intimacy with whom was not at first understood, after in vain attempting by smooth words and fair promises to induce her to go back with him, left the ship in a rage, and we were not sorry to get rid of so impudent and troublesome a visitor as he had become. Previous to leaving, he had threatened that, should he or any of his friends ever catch his faithless spouse on shore, they would take off her head to carry back with them to Múralug."[35]

David Moore, the splendid anthropologist/editor who published Brierly's manuscript notes in 1979, believes that MacGillivray misinterpreted Boroto, but he gives no good reason for this assumption. True, Barbara at one point denied having a Kaurareg "husband," though she was likely using this term in the formal sense of a spouse sanctioned by the clan, and her statement did not therefore preclude her having been Boroto's long-term lover. Technically she couldn't marry within the clan because in the Torres Strait this seems to have been a privilege denied to *marki*s (as attested by Weenie). If Barbara and Boroto had formed a sexual relationship, then it could only have been an unorthodox one, similar to the situation of Weenie on Badu.[36]

Other evidence reinforces MacGillivray's claim about Barbara's relationship with Boroto. The young warrior was a frequent topic of her conversations with Brierly, often in a way that implied intimacy. She certainly spent more time in Boroto's company than with any other young man. In Brierly's transcripts, it is noticeable that she sometimes adopted an ironic, dare one say spousal, tone when recounting some of Boroto's antics: how, for example, he'd tried to prevent her making contact with an earlier visitation of white traders, how he'd accused her of fatally weakening his European clasp knife by showing that it could be closed and opened, and how he'd passed out on Evans Bay after inhaling European tobacco into his "stomach."[37]

But Boroto was also described both by Barbara and others as physically powerful and accomplished. He was, Brierly said, quite "a wag," with the gift of the gab and a penchant for making lecherous jokes. At the same time neither Barbara nor Brierly ascribed to him any native wife or child—omissions all the odder because Boroto was a formidably influential and wealthy figure within the clan. The fact that Giom lived with Boroto's older brother Gunage and his wife, Urdzanna, rather than with her own father's family, is also significant. Both brothers were big men in the clan, and it was they who led the raiding party against the mainland Gumakudin who'd murdered and mutilated their father.

Boroto, an esteemed turtle hunter with prized expertise in making magic to entice green turtle to the surface, was also a skillful boatman, a successful trader, and one of only three men on Muralag who owned a yam garden. Despite his youth he was held in high regard by the elders of the Kaurareg, as was indicated by his key role in conducting mortuary rituals, and still more by his status as one of the clan's three feared *mydallager*s. Concern to not offend Boroto could have been one of the reasons that the Kaurareg were so reluctant to allow Barbara to return to her people. Being a *mydallager*, she claimed, gave Boroto the ability to curse and kill anyone he chose with impunity—"then when the body is found there will be no inquiry. They will only say, 'It's the mydallager.'" Fortunately for her, she never actually saw him exercise this alarming power.[38]

Barbara also told Brierly several times that Boroto had shortly before become entangled in an affair with an older widow called Yurie, a woman who, according to Barbara, had conducted a long, sly campaign to seduce him. Mutual jealousy seems to have caused a simmering feud between the two women, which flared into open conflict on at least one occasion. Barbara gave a long, exultant account of having thrashed the predatory widow in a fight. One day, for no apparent reason, Yurie had suddenly thrown a large shell filled with water at her when she was cooking.

"I ran after her," Barbara recounted, "and, as she stooped down to pick up a stick, I caught her by the hair from behind and struck her about the face. She could not do anything at all with her hands, only cried out, '*Giom, warmera* [let go], *Giom warmera*.' None of the people took her part, but they called out to me *Giom perkee*—'Strike, Giom, Strike.' They said I was a stranger among them, and the woman should not hurt me."

The fact that even Yurie's daughter supported Barbara on this occasion, and that the castaway's father, Peaqui, had to be restrained from dashing into the fray and removing the widow's head with his freshly sharpened beheading knife, suggests that the clan believed Barbara to have right very much on her side. Perhaps her eagerness to join the *marki* ship was also prompted, at least in part, by some bitterness at her partner's infidelity.[39]

Finally, there is a curious story that Barbara told Brierly about a child, to whom she ascribed no specific parents. A clan elder named Qui Qui had the responsibility of naming newborn babies, basing this on "trees or birds or anything he fancies." One particular name was exceptional enough to prompt Brierly to comment: "He [Qui Qui] must be an old wag in his way, to judge by the following name; first child—Outzie = muddy water." Brierly, unusually for him, did not name the parents of this mysterious "first" child either.[40]

We can well imagine why a young Scottish woman in that era, only eighteen years old and on the brink of returning to her family in Sydney, would want to suppress any possible suggestion that she'd been living for five years as a black man's lover, and perhaps given birth to a child by him. To admit as much would have generated a scandal to top the persisting storm over Eliza Fraser's six weeks with the Kabi Kabi people. Barbara's relationship with Boroto would have made her a social pariah in Sydney, or at the very least an object of prurient fascination.[41]

It would not be surprising if, during her five months of interviews on the voyage home, Barbara had at some point decided to take the kindly Brierly into her confidence, begging his silence. Or perhaps Brierly, in his chivalric way, guessed at her relationship with Boroto but avoided asking or recording anything that might compromise his vulnerable young informant. It's even possible that, thanks to Boroto's shipboard antics, all the senior officers of the *Rattlesnake* were aware of the relationship and chose to keep a tactful silence—all, that is, but the notoriously tactless Jock MacGillivray.[42]

A pact of confidentiality between Brierly and Barbara would partly explain another puzzle: why the redemptive story of her life with the Kaurareg Islanders of the Barrier Reef sank into immediate oblivion. By the time the *Rattlesnake* berthed in Sydney in early February 1850, Barbara had recovered both her mastery of English and her health. She was, MacGillivray said, "handed over to her parents . . . in excellent condition." The *Brisbane Courier* had published a short account of the story even before the *Rattlesnake* completed its voyage, but the *Sydney Morning Herald*, which might have been expected to make much more of the sensational story, produced only a cryptic factual description of the young castaway's five-year sojourn with the Kaurareg.[43]

Sanitized in this way, the young working-class girl's story probably appeared too positive for many colonial tastes and too boring to be newsworthy to other papers. Perhaps, too, Captain Owen Stanley, Oswald Brierly, and the other European officers of the *Rattlesnake* managed to exercise some kind of muting influence on the Sydney paper, thus protecting Barbara. But if they did, it was at the expense of providing a significant counterweight to the burgeoning myth of savagery among the peoples of the Barrier Reef.

The absence of any sexual sensation in her story helped Barbara Thompson slip into obscurity. Twenty-one months after returning to Sydney, she married a sailor, James Adams, who seems to have died within a few years. In 1876 she married another sailor, John Simpson. Barbara eventually died in Sydney in 1916, at the age of eighty-five, and is buried in Rookwood Cemetery in Sydney. Boroto probably died in 1869 on Muralag, after a brutal white reprisal conducted in error against the Kaurareg. These mass killings led to the near extinction of his Kaurareg clan.[44]

Brierly's collection of illustrated journals remained unpublished in his lifetime and for long afterward, thereby making no public impact in his day. Instead Brierly sold his material to a local book collector, David Mitchell, who became the founding patron of Sydney's Mitchell Library, thus ensuring that this story of *marki* love on the Reef survived the annihilation of the Kaurareg clan who'd made it possible.

The climate of racial opinion in colonial Australia and in Britain was hardly conducive to publishing the testimony of a castaway so at odds with the titillating traumas of Eliza Fraser. Brierly's continuing reluctance to expose Barbara Thompson to public scrutiny might have been another thing constraining him from publishing his journals, and might explain, too, why his many sketches and paintings of the Kaurareg contain no representation of Barbara herself. Indeed, no portrait of the young castaway seems to have survived anywhere.

But perhaps another reason Brierly failed to publish was because he never found time to work up his material. Soon after docking in Sydney, he was hired as artist to the British survey ship HMS *Meander,* after which he served as naval artist to the British Fleet throughout the Crimean War (1853–56). With the coming of peace, he worked briefly on Queen Victoria's royal yacht, before joining a series of imperial cruises in the 1860s as marine artist to the Duke of Edinburgh and the Prince of Wales, one of which brought him back to Australia. Oswald Brierly died in England in 1894, after having been knighted for his work with the Royal Yacht Squadron.

It was all a far cry from squatting in the sand with his Kaurareg friends. He had found, on his return to Australia in the mid-1860s, that anti-Aboriginal sentiment had hardened appreciably. The northward expansion of European settlement into New South Wales and Queensland was by now generating a virulent propaganda against Aboriginal and Torres Strait Islander peoples, to justify the dispossession of their lands and resources. The two young castaways who were soon to follow in Barbara Thompson's wake would find reentering white society a much more painful affair.

6

HEARTLANDS

The Lost Lives of Karkynjib and Anco

L ONDON IN THE FINAL YEARS of the nineteenth century had lost none
of its appetite for sensationalism, and in August 1898 a new purveyor
of "true-life adventure stories," *Wide World Magazine*, began a twelve-
month serial. The hyperbolic claims of the title—"The Adventures of
Louis de Rougemont. Being a Narrative of the Most Amazing Experiences a Man Ever Lived to Tell"—did not disappoint, and the work soon
became a bestselling sensation.

De Rougemont, a tall Swiss sailor with a seamed brown face and
ready tongue, told a tale that made Robinson Crusoe's sound prosaic.
Shipwrecked on a reef north of Australia while sailing a pearl lugger, he
was marooned for several years on a tiny strip of sand. With great inge-
nuity and courage, he claimed, he subsisted on raw fish stolen from the
local pelicans, and built a hut from the piles of pearl-filled oyster shells
lying on the reefs. During his long marooning he'd discovered some
amazing properties of turtles: he could ride on their backs while steering
with his toes, he could germinate plants in their blood-filled shells, and
he could dive underwater with them to discover a marine Eden.[1]

Rescued by three Australian natives, he was taken to the world's last
great wild frontier, somewhere in northern Australia. Here, he maintained, he "discovered" a vast area of lush virgin wilderness where jewels

and gold nuggets lay discarded for the taking, and where his genius earned the admiration of cannibal chiefs and native women.

That de Rougemont was an exaggerator will be obvious, but he was in fact a serial fantasist who would end his life performing at fairs under the title of "The Greatest Liar on Earth." Born Henri Louis Grin on November 12, 1847, to a peasant-farmer family in Gressy, Switzerland, he migrated to Australia in 1875 to work as a butler for the new governor of Western Australia. Once there, he left his employ to live the life of a drifter and con man for twenty years, having enough real-life adventures to ensure that his later castaway tale was half plausible. During that time he married Eliza Jane Ravenscroft of Newtown, Sydney, and fathered seven children.

While living in Western Australia in 1875, he acquired a small, deformed-looking cutter known on the Fremantle waterside as *The Sudden Jerk*. After fitting it out as a pearling lugger, Grin recruited some white riffraff and a few press-ganged Aboriginal divers to cruise with him for several years on the northwest coast. There he somehow managed to fail in the pearl-shell trade at a time when others were making fortunes.

Eventually he and a brutish white associate became implicated in the murder of an Aboriginal diver. Grin escaped the perfunctory notice of law officers by sailing around the top of Australia and down through the Great Barrier Reef lagoon to Cooktown, where he claimed to be the sole survivor of an Aboriginal attack. Soon after, he joined the Palmer River gold rush, inland from Cooktown, before mounting at least one further pearling expedition northward. This ended abruptly on a coral reef off the Torres Strait in 1880. Two years later Grin appeared in Sydney, where he pecked a living as a dishwasher, real estate salesman, photographer, and marketer of a patented diving suit that killed its first demonstrator. In 1897 he abandoned his Sydney family for a brief stint in New Zealand as a spiritualist, before working his passage to England.[2]

Grin arrived in London with a memoir in mind. Deciding to season his adventures with a little historical fact, he trawled through a book in the British Museum called the *Australian Dictionary of Dates and Men of the Time*. Compiled by a former colonial journalist, it contained true-life castaway stories of sailors who'd lived with Aboriginal clans. Two in particular grabbed Grin's attention, both young men who mid-century had spent seventeen years living with separate groups in the region of the

Great Barrier Reef. James Morrill, an English sailor of twenty-two, had been shipwrecked in 1846 and rescued by a clan of Birri-Gubba speakers near present-day Townsville. Narcisse Pelletier, a French cabin boy of fourteen, was rescued from a shipwreck in 1858 by a group of Wanthaala near Cape Direction, midway down the Cape York Peninsula.

Though he plundered both stories, Grin found Pelletier's the more enticing. Being a French speaker, Grin had little difficulty substituting himself for the cabin boy, though he chose to award himself a more aristocratic lineage. Thus was Louis de Rougemont, north Australian castaway, born. Had he been able to control his rampant imagination enough to steal only the factual details of the two stories, Grin might still have created a sensation in London without committing the blunders that eventually exposed him as a fraud.

For Grin had chanced on something rare: two accounts of the lives and habitats of hunter-gatherer clans in the Reef region on the verge of the engulfing experience of European contact. Both stories were rich in the Crusoe-like details that Grin itched to borrow. He decided, however, to excise their uncomfortable endings. The two boys, along with the clans and estates that nurtured them, had eventually become casualties of the predatory frontier invasions that Grin himself had helped to mount.

These are the two Reef castaway stories that *Wide World Magazine* ought to have carried.

Although Jem Murrells (later James Morrill) and Narcisse Pelletier were born twenty years apart—in 1824 and 1844 respectively—and in the continentally separated villages of Abridge in Essex and Saint-Gilles-sur-Vie in the Vendée, they were carried to the Great Barrier Reef by the same flow of European trade. Both were from artisan families living in villages linked to a rising seaborne imperial commerce centered on the present-day Asia-Pacific region.

The sailing barges of Abridge, still famous today, plied the Blackwater River a short distance to Maldon, a port where seagoing colliers and merchant trading ships abounded. Morrill's father, a millwright, allowed his son a brief education at the local Church of England elementary school before making him join the family workshop at the age of fourteen. At a time when "the fine white sails, and the beautiful sea quite charmed me—I was always wishing I could be a sailor," Jem one day impulsively signed on as a cabin boy with a Maldon collier, the *Royal Sailor*, a ship on which he eventually completed a full apprenticeship.

Around 1845, again seeking wider pastures, Morrill joined the crew of the *Ramalees*, which was carrying troops to Hobart, Van Diemen's Land (present-day Tasmania), on the other side of the world. After some local voyaging in the South Seas, he set sail from Sydney in February 1846 to return to England on Captain George Pitkethly's Dundee-based merchantman, the *Peruvian*, which was following a standard colonial pattern by carrying a cargo of hardwood to China en route for home. Three nights later, caught in heavy seas, the ship hit the Horseshoe Reef, one of a deep-sea chain at the southern end of the Great Barrier.[3]

When the twenty-two-year-old able seaman woke on the splintered deck at daybreak on February 28, 1846, "a terrible scene presented itself, as far as the eye could reach there were the points of the rocks awash, but no friendly land in view." Having already lost several men when the cutter and jolly boat were pulverized on the rocks, pious Captain Pitkethly gathered the survivors to pray, and then ordered them to build a raft from masts and spars. Twenty-one surviving "souls," including several women and children, clambered onto this precarious craft, along with a little keg of water, a few tins of preserved meat, and some brandy. Promising not to eat one another should they run out of food, each agreed to a daily ration of one tablespoon of preserved meat and four measured sips of water.[4]

Supplementing this fare with fresh blood and raw meat from the occasional seabird, they drifted for twenty-three days before the older men and the women and children began to die from hunger and thirst. The remaining sailors, who had fishing hooks but no lines, improvised a snare using an oar baited with a freshly severed human leg. By this means they caught and ate several of the voracious sharks that continually circled their raft. When the current eventually crashed them onto the Great Barrier Reef, they could barely drag the raft across the coral into the lagoon.

Two or three days later they beached on the southern point of Cape Cleveland, near modern-day Townsville. Too weak to crawl in search of food and water, several sailors died almost immediately. Four survivors— Morrill, an unnamed cabin boy, and Captain and Mrs. Pitkethly— managed to subsist for a further ten days on rock oysters and rain pools, before they heard Mrs. Pitkethly suddenly cry out one evening, "Oh George, we have come to our last now, here are such a lot of the wild blacks." Twenty or thirty naked warriors had emerged from the scrub and were staring at them edgily.[5]

Morrill braced himself while the warriors prodded him and his three emaciated companions to find out whether they were truly human. Apparently satisfied, the men "took pity on us." Half a dozen older warriors lay down beside them in a nearby cave; the remainder brought them delicious, nutlike roots to eat. Too weak to join in a celebratory corroboree, Morrill croaked a heartfelt rendition of some lines from a half-remembered hymn, "Light Shining Out of Darkness."

Prior to moving off to a small nearby camp, some members of an inland clan from Mount Elliot (the Bindal) formally claimed Morrill and the cabin boy, while others from a coastal Cape Cleveland clan adopted the Pitkethlys. One warrior carried the half-dead cabin boy; others supported the three hobbling adults. Once they were at the camp, two or three senior men gently reassured Morrill by pressing their fire-warmed hands against his shivering body.[6]

A few days later the castaways were moved again, some five miles farther inland to a more permanent camp of around fifty people. Here they were given a *gunyah* (hut) to sleep in and plenty of food and drink. With the onset of the dry season their hosts decided to settle at this place for several months, to stage what Morrill called a *"boree"*: the ritual of initiating young males into manhood undertaken at gatherings of local clans. The rescuers also took the opportunity to perform a succession of corroborees to tell the strange story of the castaways to the incoming clans, who eventually numbered a thousand individuals.

During this period, the four castaways picked up a smattering of Birri-Gubba language and food-gathering practices. It was probably around this time, too, that Morrill was given the name Karkynjib Wombil Moony, after one of the Aboriginal elders. "We spent our time in wandering about with them on their fishing excursions," he recalled, "and in learning to snare ducks, wild turkeys, geese, and other wild fowls, which I became very expert in after a while." At the completion of the initiations, when the clans began returning to their own districts, the castaways asked permission to attach themselves to southerly based clans, which were likely to be situated closer to European settlements. For two further years, Morrill thus lived with a clan of Jura, or Gia, people based around Port Denison, before a quarrel induced him to rejoin the Mount Elliot Bindal.[7]

Back in the bustling fishing and trading port of Saint-Gilles-sur-Vie, Narcisse Pelletier, too, had refused to follow his father's profession, preferring the more exotic maritime traditions of his mother's family to the tedium of working as a village shoemaker. After an even sparser schooling than Morrill's, he made his first sea voyage at the age of eight on a boat owned by his uncle. At thirteen, having undertaken three further voyages as a cabin boy, he was badly wounded by the first mate of the merchantman *Reine des Mers*. Whether there was a sexual element to this attack remains unknown, but it is possible: Pelletier was much later diagnosed with "venereal testicles."[8]

Departing the ship at Marseille, he joined the merchantman *Saint-Paul* under the command of Captain Emmanuel Pinard, who was engaged in the same Asia–Australia trade as Pitkethly. After carrying French wine to Bombay, the *Saint-Paul* called at Hong Kong around August 1858 to load a human cargo of 317 Chinese laborers bound for the New South Wales goldfields. Having skimped on rations for the long journey, Pinard decided to risk a dangerous shortcut between New Guinea and the Great Barrier Reef. Caught in a storm, the ship hit a coral reef off Rossel Island in the Louisiade Archipelago, 125 miles southeast of New Guinea.[9]

Before their ship broke up, the crew and passengers managed to reach a tiny waterless strip of rock and guano—presumably the inspiration for de Rougemont's fictive marooning. Across a shallow strait stood the picturesque wooded island of Rossel, inhabited by Melanesian tribesmen. A small contingent of sailors who waded across to obtain fresh water were first welcomed but then later attacked with arrows, clubs, and stones. Many of the sailors were killed, but the fourteen-year-old cabin boy Pelletier managed to escape, nursing a severe head wound.

After repelling a further onslaught, the captain and eight or so crewmen secretly decided to sail the longboat to Australia for help, leaving their Chinese passengers on the island with rifles and a few provisions. Guessing their plan, Pelletier managed to swim out and intercept the longboat as it was leaving. Had he not done so, he, too, would have been killed and eaten, along with the bulk of the marooned Chinese.[10]

For the next twelve days the longboat drifted in light winds and scorching heat. The crew survived by eating a few seabirds, drinking their own urine, and catching occasional mouthfuls of rainwater in their boots. Weaving their way through a Reef channel, they eventually made landfall somewhere near Cape Direction. Here they found a native well, which the adult sailors drank dry. While waiting for the water to replenish, Pelletier fell asleep, returning later to the beach to find the longboat gone. Near dead from blood loss, starvation, and thirst, he woke the next morning to glimpse three naked black women scurrying into the bush. Half an hour later, he faced two spear-carrying Aboriginal warriors, one of them "horrible to behold."[11]

A Nantes scholar and medical man called Constant Merland, who later told Pelletier's story in a French publication, asked readers to imagine the cabin boy's plight, alone on the wild coast of "Endeavour Land" (Cape York Peninsula): "We can understand the anguish he must have felt when we reflect on the scenario he had before him. To die of hunger and thirst, to become the prey of fierce beasts or to be eaten by the savages, such were the dreadful alternatives that seemed to be in store for the poor child."[12]

Facing the two warriors, the wounded boy held out a small tin cup and a handkerchief as overtures of goodwill. The men in turn held out their hands in reassurance, offered him water and fruit, and supported him to where their wives were seated around a fire. Falling into an exhausted sleep, Pelletier woke the next morning to find that they, too, had gone.

Despairing at having again been abandoned, the boy was overcome with joy when the warriors suddenly reappeared, carrying breakfast. "[H]e and the two men eagerly rushed to greet each other," a reunion capped by his reciprocal gift of some ship's blankets that produced "shouts of joy and . . . the most vigorous displays of friendship and affection." This bond was to prove permanent: one of the two warriors, Maademan, having no children, decided to adopt the boy. He called him by a name that Constant Merland transcribed as "Amglo" or "Anco."[13]

Pelletier was taken to a camp of thirty to fifty individuals, the temporary headquarters of a small maritime clan of Wanthaala people who spoke the Uutaalnganu language—the same group whose members had in 1843 killed a man in Joseph Beete Jukes's survey party at Cape Direction. Here Anco was introduced to his future kin and companions, and "little by little . . . he took on all the ways of the people with whom he was living. After a certain time all that distinguished him from them was the color of his skin and the shirt and trousers which covered his body. It was not long before this last feature disappeared . . . So there he was, he, too, in the primitive state."[14]

Despite both shipwrecked boys being rescued within the Reef region, the estates of their clans were 620 miles apart. Morrill roamed, fished, and hunted very widely around the Burdekin and Herbert rivers region, but his primary clan was located inland, eighteen to thirty miles southwest of modern-day Townsville, around Mount Elliot. This heartland was rich and varied, a mix of steep mountainsides and deep ravines studded with patches of high-growing open eucalyptus forest and dense lower reaches of rainforest, all of it intersected by fast-running streams that fed into freshwater lagoons. During the dry season, Morrill also spent periods camped at his clan's regular fishing spots on the Burdekin River and among the Cape Cleveland coastal swamps. In between these two zones of mountain and coast, he hunted on stretches of grassy open plain, created, as the explorer Augustus Gregory observed in 1856, by the Aborigines' regular firing of scrubby undergrowth in order to promote grasses on which kangaroo and wallaby would feed.[15]

Such biodiversity meant abundant food resources in both dry and wet seasons, despite inevitable periods of drought and cyclone. Morrill was particularly proud of his skill in weaving and setting string snares and wild flax nets for capturing geese, wild fowl, and duck, as well as fish and wallaby. His published memoir also mentions collecting breadfruit, procuring honey from the tree hives of native bees, and, in his early years, digging regularly for yams and roots. He describes the small children of the clan "setting roots" in the swamps, presumably so they would sprout the following season—surely a mode of farming by anyone's standards.[16]

His diet also encompassed shark, alligator, shrimp, shellfish, kangaroo, rat, snake, grubs, snails, pigeon, and turkey. On top of all this protein, his vegetable intake included three "delicious" root staples, one mountainous, the others scrub-based; two small, turniplike roots that grew in open grass; a leafy riverbank creeper; another, smaller creeper with a turniplike root; a blue-flowered creeper which ran among grass; "and many more or less like them." Along with breadfruit, common fruits included a native plum; blue, white, and red native currants; a wild banana; a wild apple; and a red and a black fig.[17]

Even so, managing this habitat was not for the fainthearted, as is suggested by the fact that Morrill's three fellow castaways died from natural causes within a few years of their adoption. Morrill was burly and strong, like "the fine race of people" that adopted him. Still, he was unable to avoid being bitten by a crocodile, which damaged his knee permanently, and by a whip snake, which caused him to swell up for several days. These, he said, were commonplace occurrences: he'd seen "dozens" of natives taken by saltwater and freshwater crocodiles, or bitten by venomous snakes, though he didn't mention whether his Bindal and Jura companions shared his affliction of acute rheumatism, the result of lying for long periods on hard wet surfaces. When he first returned to the European world, Morrill was described as having "holes in his forehead, arms and body, and marks of sores and scurvy."[18]

Pelletier's "country," a coastal estate, was situated halfway along the Cape York Peninsula, not far from the modern Aboriginal settlement of Lockhart River, a hotter, wetter, and more isolated region than Morrill's. This area, within which Pelletier knew every rock, tree, and bush, was for him also animated by the "mythic charter" of the clan's spiritual origins. Known today as Sandbeach country, it consisted of a belt of white, sandy, dune-filled coastline, riverine mangroves, and fringing scrub. North to south it extended about seven miles, then twenty-five miles eastward across the sea to the edge of the Barrier Reef, and another six miles or so inland through scrub and grassland to the foothills of a rainforested mountain chain.[19]

Beyond this small family estate, the larger influence of Uutaalnganu-speaking peoples probably stretched for some forty-odd miles, from Lloyd Bay at the mouth of the Lockhart River to Cape Sidmouth in the south. At its midpoint lay a small, resource-rich, and sacred offshore cay called Night Island.

Modern anthropological authority Athol Chase has called this region

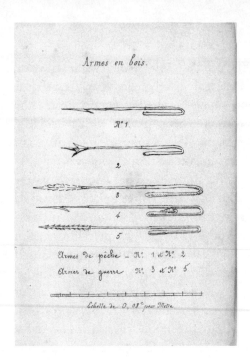

Armes en bois by Narcisse Pelletier. In *Dix-sept ans chez les sauvages: Narcisse Pelletier* by Constant Merland, 1876 (Mitchell Library, State Library of New South Wales)

A Sandbeach man and canoe in the 1930s (Lockhart Images 1930, John Oxley Library, State Library of Queensland)

"possibly one of the richest and most varied environments for hunter gatherers anywhere in the world." Pelletier described to Constant Merland a way of life that included hunting on land, often with the aid of dogs and nets, for snakes, crocodiles, lizards, echidnas, parrots, cockatoos, wild fowl, pigeons, cassowaries, and emus. Women and children of the clan developed specialized skills in collecting eggs of all kinds, including those of megapodes, turtles, and crocodiles. They also gathered some sixty or seventy fruits in different seasons, greens, yams, edible mangroves, and a dozen or so other vegetable staples, as well as a large range of estuarine and reef shellfish.[20]

Firing the scrub was the method male hunters favored to drive animals into ambushes, and entice them to graze on freshly cleared and sprouted grasslands. On this subject, even Merland, who shared the opinion of most Europeans that Aboriginal people lacked any concept of cultivating or managing crops, made a significant exception: "unless from the agricultural viewpoint, we wish to consider the care the savages take in firing the woods where the yams grow so that the tubers of these plants develop more extensively and their crop is more plentiful." He was describing what Australian ecologists now call firestick farming.[21]

Yet for Wanthaala menfolk like Anco, all of this was subsidiary to hunting and fishing in the Barrier Reef lagoon. They paddled as far as twenty-five miles out to sea in small outrigger canoes to harpoon the clan's most prized foodstuffs: huge, shy dugong, four species of turtle, some fourteen species of ray, and a variety of large meaty fish like rockfish and grouper. Such activities were viewed as the most dangerous, skillful, and desirable of all male pursuits. Thanks to being a *pama watayichi*, or dugong man, Anco became "quite a personage in his tribe." He owned his own boat, was a famed manufacturer of ropes, harpoons, spears, and spear throwers, and probably worked as a skilled shipbuilder. Certainly he impressed Merland with his account of how the Wanthaala built seagoing craft by felling trees, cutting planks, and hollowing out logs with saws and blades made by grinding the edges of shells and stones.[22]

Aside from the threats posed by sharks and saltwater crocodiles, sudden storms made these craft vulnerable to capsizing, something that happened to Anco and his inseparable "cousin" Sassy when hunting green turtle after dark. Both were strong swimmers, but had to spend harrowing hours clinging to their detached outrigger floats until a lull in the storm enabled them to make it to the beach, where the community waited with lighted torches.[23]

Sharing out portions of turtle and dugong according to protocols of clan seniority could also prove hazardous: Anco seems several times to have caused offense on this score. Once he was saved from a spearing by Sassy's intervention; on another occasion he incurred a formal curse that was said to have produced an incurable ulcer on his leg.[24]

In 1848, after the last of his fellow castaways had died, James Morrill was for a time the only white man for some two million square miles. A decade later all this had changed. The founding of the independent state of Queensland in 1859 opened up an area for potential settlement two and a half times the size of Texas. Glowing descriptions from explorers of what was formally named the Kennedy District, based on the drainage areas of the Burdekin and Herbert rivers, soon reached would-be settlers in Britain and New South Wales. The region's soils, rivers, and grassy open plains—created by Aboriginal fire regimes—looked ideal for sheep and cattle.

Thanks especially to the efforts of the entrepreneur and explorer George Elphinstone Dalrymple, vast new pastoral runs were opened up to squatters from about 1860, many of them Scots aristocrat families like Dalrymple's own. Instant port towns mushroomed at Bowen and Mackay in support. By 1863 almost the entire Kennedy District had been taken up, and squatters were beginning to push northwest into the Gulf Country.[25]

The seasonal hunter-gatherer range of James Morrill's clan lay at the epicenter of this invasion. In a terse description of its impact on his clan relatives and friends, Morrill marks the beginning of one of northern Australia's most intense and sustained bouts of frontier conflict. Naive "good intentions" on the part of some squatters like Dalrymple couldn't soften the fact that the Aboriginal clans of the Burdekin–Herbert district experienced a full-scale assault on their estates, ecologies, and cultures. Although the Land Act of 1860 technically gave Aborigines the right to enter leased land, there were many ways to keep them out, including guns. Squatters appropriated the rich estuaries, swamps, rivers, lagoons, forests, and grasslands that the clans had inhabited for centuries, and then the same invaders literally locked the clans out of the new European economies that sprang up in their place—the pastoral runs, stations, townships, and ports. Armed resistance from Aborigines led in turn to a hardening policy of forced dispersals and shootings at the hands of specially formed troops of black police and squatter vigilantes. [26]

Morrill, initially excited at news of the arrival of his countrymen in the district, was then troubled at the ensuing treatment of his kinfolk. In 1860 an attempt by "a stout able-bodied blackfellow, a friend of mine" to tell Morrill's story to the visiting survey ship *Spitfire* ended when the Aboriginal man's actions were misinterpreted and he "was shot dead . . . and another was wounded." Soon after this, Morrill learned that the funeral ceremony of a respected clan elder had been interrupted by a gunshot from a squatter, which killed the man's lamenting son. On top of this, cattle were reported to be milling at the Bowen River "in great numbers," drinking the water holes dry and leaving the fish to die. Worst of all, fifteen men in "a fishing party belonging to the tribe I was living with, were shot down dead," probably by native police.[27]

Despite the accumulating carnage, Morrill could not suppress his "hopes of being restored." Eventually he persuaded his troubled clan to let him visit the white intruders in order to negotiate an end to the bloodshed and spoliation. Yet when the time to depart finally came, his heart was torn: "They then said you will forget us altogether; and when I was coming away the man I was living with burst out crying, so did his gin, and several of the other gins and men. It was a touching scene. The remembrance of their past kindness came full upon me and quite overpowered me. There was a short struggle between the feeling of love I had for my old friends and companions and the desire once more to live a civilized life."[28]

On Sunday January 25, 1863, a bronzed, naked, and shaggy Morrill hailed two stockmen named Wilson and Hatch at Sheepstation Creek, and only narrowly averted the fate of his clan relatives by yelling out in broken English, "Do not shoot me, I am a British object."[29]

When Narcisse Pelletier made his contact with Europeans in 1875, he did so under markedly different circumstances. A small but growing frontier industry of English and Japanese trepang and pearl-shell luggers was taking over a long-standing trade with China that had previously been run by seagoing Malays. Now based in the Torres Strait, it was beginning to push southward into the rich coastal waters of Cape York. During the 1870s, marine-resource hunters eventually reached the remote waters of Sandbeach country, the territory of Pelletier's clan, and as Louis Grin had done earlier, they were initiating contact with remote Aboriginal communities in order to lure or compel young men into working as divers.[30]

In early April 1875 the *John Bell*, a brig belonging to a Scots entrepreneur in the Torres Strait called Joseph Frazer, anchored off Night Island in search of fresh drinking water. Prior to this, Pelletier's clan had probably experienced only one European incursion into their estate. In July/August 1860 the former *Rattlesnake* naturalist Jock MacGillivray had arrived on Night Island aboard a commercial trader, searching for sandalwood to trade with China. Looking toward the mainland beach, Mac-

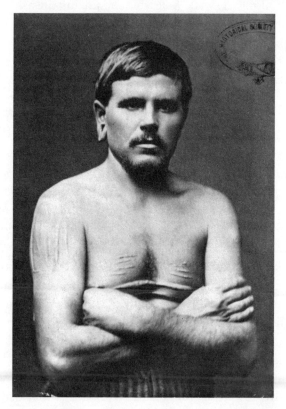

Narcisse Pelletier, c. 1875 (Royal Historical Society of Queensland
photographic collection)

Gillivray glimpsed what he thought might be a young half-caste, but he was unable to get close enough to confirm the sighting. Had the Wanthaala not hidden young Anco from view, the naturalist might have featured in the discovery of another castaway.[31]

Pelletier's eventual reentry into European society was coerced. On April 11, 1875, the *John Bell*'s watering party noticed a wild-looking white man, complete with a wooden earlobe plug and chest scarifications, moving among a group of Aborigines. Without hesitation or consultation they "rescued" him. Pelletier, unable to speak a word of English, dared not resist his "kidnapping" for fear of being shot. His Wanthaala relatives, who, *John Bell*'s captain said, "were very reluctant to part with him," urged Anco to jump ashore, but he was forcibly prevented.[32]

A fisherman at Somerset, the small government settlement on the tip of Cape York where Pelletier was taken, reported that the castaway had been "fastened so that he could not escape." Other observers remembered him behaving like a caged bird: he crouched on his heels, "casting quick, eager, suspicious glances around him on every side and at every object which came within his view, rarely speaking, and apparently unable to remember more than a few words of his own language."[33]

As Constant Merland would later declare, Pelletier "was no longer a Frenchman, he was an Australian." Adopted at such a young age and then living for so long with the Wanthaala, "his naturalization was complete." A variety of French speakers tried to question him soon after his capture, first at Somerset, then during his voyage south on the *Brisbane*, and again while he was staying in Sydney prior to returning to France. All noted his limited recollection of both the French language and his original family. George Eugène Simon, the Sydney-based French consul, captured something of the boy's pain as he struggled to recover a lost past:

> I told him the name of his village and I then witnessed some of the strangest and most painful spectacles, I think, that one might see: this wretched man made extraordinary efforts to remember; he wanted to speak to me and all that came to his lips were inarticulate sounds. . . . His face and eyes expressed a terrible anxiety and anguish, and something like despair which was painful. I suffered with him and almost as much as him. Sweat was breaking out on my brow as on his. Involuntarily I remembered the tale of Hoffman's of the man who has lost his

shadow and his image. I would have done anything to give him back then and there his identity, which clearly he was trying to grasp hold of again.[34]

Naturally enough, the two young castaways faced different challenges when attempting to rejoin European society. Because Morrill rejoined his countrymen voluntarily, his sense of dislocation was initially less severe than Pelletier's; moreover, he was older and had retained more of his original language and culture. He made a symbolic renewal of his ties to Protestant Christianity by getting himself rebaptized, but even so, he found himself in the precarious situation of having to reintegrate in the midst of a brutal frontier war.[35]

After being welcomed as a local hero in the nearby port town of Bowen, Morrill was taken to Brisbane, where he delivered a short version of his story to a local *Courier* journalist, Edmund Gregory, who republished it as a pamphlet. From the outset, though, Morrill showed signs of unease when discussing his former life, as he tried to balance the pressure to affirm white prejudices with his private hopes of stopping the settler attacks on his people. Although the governor of Queensland, Sir George Bowen, thought him "intelligent and respectable" and arranged for him to receive a plot of land in Bowen and a government position as a customs officer, Morrill's interview in Rockhampton with the trigger-happy police magistrate John Jardine proved less comfortable. Nevertheless, a year later Morrill was sufficiently assimilated to marry a young immigrant domestic servant, Eliza Ann Ross, who soon gave birth to their son.[36]

Some settlers saw Morrill as a useful asset because of his mastery of Aboriginal language and his intimate knowledge of local geography, animals, plants, and bush medicines. A few also encouraged him to fulfill his desired role of operating as mediator with the beleaguered clans of the district. When first meeting him at Sheepstation Creek in 1863, Wilson and Hatch had immediately asked him to tell the Bindal people that "if they did not interfere with us, we should not interfere with them." Morrill's interpretation of the message was actually a good deal bleaker and more realistic: "I told them that the white men had come to take their land away. They always understood that might, not right, is the law of the world, but they told me to ask the white men to let them have all the ground to the north of the Burdekin, and to let them fish in the rivers; also the low grounds, they live on to get their roots—ground which is no good to white people, near sea coast and swampy."[37]

Both his clan's and his own request of a reward for "the natives who were so kind to me" were ignored. True, one of his more sensitive confidants, George Dalrymple, decided to use Morrill as a negotiator when setting up a new township at Rockingham Bay, north of Bowen, in 1864. Morrill was taken aboard the schooner the *Policeman* and asked to jump into waist-deep water to relay a message to a band of muscular warriors who'd waded out to meet them. As Dalrymple recalled:

> I told them, through Morrill, that we had come to take possession of the coast from a point on the northwest shore of the bay to a point opposite Haycock Island, and that we were going to settle there and possess it.
>
> They said "they hoped we were not going to war with them." I replied, "No: that we did not wish to hurt them, but that we wished to be left alone; that if they would keep off and not molest us, we would not injure or interfere with them in any way." They seemed to understand this ultimatum, and retired slowly into the mangroves; Morrill having explained it to them over and over again, and told them to inform the neighbouring tribes accordingly.[38]

Out of these "negotiations" came the Reef township of Cardwell. But the fate of a go-between in such vexed circumstances is usually to please neither side, and so it proved with James Morrill. His Aboriginal friends felt betrayed, their pleas for access to marginal land having been ignored, and his offers to help negotiate peace were rejected because settlers suspected him of conspiring with black resisters. *The Courier* suggested that "wave-like" successions of black outrages against white settlers flew in the face of Morrill's claims about the natives' peaceable disposition. Three months later the same paper reported on a settler who'd promised "to give [Morrill] a small piece of lead" if he tried any peacemaking on his property.[39]

Though on the surface Morrill might have appeared to make a reasonably successful transition back into white colonial society, Dalrymple believed him to be "an unhappy man, trying to reconcile the hostile suspicious attitude of the settlers to the Aboriginals, with his own loyalty to the blackfellows who had looked after him so long." As early as 1864

Morrill reached the melancholy conclusion that "the work of extinction is gradually but surely going on among the aboriginals. The tribe I was living with are far less numerous now than when I went among them." Their only hope of survival, he believed, was to hide out in inaccessible terrain like Mount Elliot: "it is such thick scrub, and there is such an abundance of food in it, and plenty of water, that if the Aboriginals were driven from the country all around they would find safe asylum there." He held destructive police and settler policies responsible for his clan's rapid population decline, and blamed the disruption of their traditional way of life for the deterioration of their physical and mental health.[40]

This depressing state of affairs, along with the debilitating effects of his rheumatism and injured knee, precipitated Morrill's early death, on October 30, 1865 at the age of forty-one, only two years and nine months after his reentry into the white world. The Bowen townsfolk staged a grand procession for his funeral, but the local newspaper makes no mention of their having "let in" any of his former kinsfolk to join the mourning.[41]

Narcisse Pelletier's return to "civilization," while different from Morrill's, was no less ambivalent. Pelletier had come to welcome the idea of going home to his French family but he also retained a pride in his former clan identity and values. He'd been deeply missed by his mother, and was joyfully received by the villagers at a celebratory bonfire outside the Pelletiers' house, but after a while his family seems to have found his "savage" appearance and pagan beliefs confrontational, if not repulsive. A decision to have him exorcised by the local Catholic priest suggests something of their unease. Local rumor had it that Pelletier came to feel a reciprocal alienation.[42]

His greatest initial challenge was to cope with public and academic curiosity about the nature of "savagery" at a time when new currents of "scientific" racial theorizing were coming into fashion. Sensitive responses like that of amateur anthropologist Louis de Kerjean in 1876 were rare. "Instead of cannibals," de Kerjean wrote, "the young sailor had found an adoptive father, whose memory will always remain dear to him, and a second homeland, which nonetheless had not made him forget either his family or the country of his birth."[43]

More typical were the writings by members of the Société d'Anthropologie de Paris, which had been established by the physician and anatomist Paul Broca at the beginning of the decade. They saw themselves as radical, secular, objective practitioners of a new science of physical anthropology, which used comparative craniological and anatomical measurement to assess what Broca called "the respective position [of races] in the human series." Suspicious of the humanistic values of sympathy and imagination associated with ethnography, most members espoused a "polygenic" theory of multiple human racial origins—as against "monogenists," who believed in a single origin. Société thinkers generally categorized Australian Aborigines as the most inferior racial type on the ladder of the human species, barely separated from animals.

Whether the French Consul George Simon, Pelletier's early questioner and patron, held such hard-line ideas is uncertain, but he was said to have been briefed about Aboriginal peoples by an eminent Société theorist, Paul Topinard. Topinard placed Anco's clan, the coastal Wanthaala, even lower on the scale of savagery than their inland counterparts.[44]

Constant Merland, the Nantes scholar who interviewed Pelletier in 1875 to produce the biography *Dix-sept ans chez les sauvages*, seems to have been a more sympathetic figure. Although he organized Pelletier's story into a scientific-style treatise, he saw himself as an ethnographer rather than a physical anthropologist. Most importantly he acknowledged that some of the Wanthaala's values were strongly indicative of the "tribe's" humanity. Maademan, he said, was a devoted adoptive father, and the boy Sassy "a true and faithful friend." Wanthaala women likewise showed intense and enduring maternal feelings toward their children. All members of the tribe were selfless in their sharing of property and impressive in their respect for the dead. Furthermore, Pelletier's linguistic evidence "proved that the vocabulary of these tribes is truly rich," even including words to express such complex concepts as a species and a genus.[45]

Reading between the lines, though, one senses that Pelletier engaged in a covert resistance to Merland's more insensitive questions and implications, including some that mirrored prevailing assumptions of Société anthropologists. As an initiated warrior, Pelletier had sworn never to disclose major realms of men's knowledge and sacral mythology, hence his silence about male initiation rituals. He also avoided mentioning his personal relationships with Aboriginal women, as did Morrill, albeit for different reasons. Later evidence suggested that both men might have left behind several children.[46]

The two former castaways were also evasive about the explosive subject of cannibalism, though Pelletier once muttered, *"ce n'est pas jolie,"* and Morrill admitted that it was practiced occasionally on slain enemies. One of Pelletier's most sympathetic early questioners noted his evasiveness: "I am inclined to think, he had definitely made up his mind to give us no more information about the tribe and the language."[47]

Merland couldn't help testing Pelletier's responses against two poles: "civilized" Frenchmen on the one hand, and the "bestial" Wanthaala, who were "living a completely animal existence," on the other. The latter, Merland claimed, were moved by coarse material and instinctual appetites: "their thought never soars toward higher realms, it never embraces intellectual questions, it never debates accepted beliefs." Almost uniquely among all "savage" peoples on earth, he speculated, they lacked a belief in immortality or a higher spiritual order. Expressions of sentiment and feeling were unknown: the men were brutal to their wives and endemically warlike toward their neighbors. All in all, he thought the Wanthaala were fixed in a stasis that blocked any will to improve. Narcisse Pelletier, Merland concluded, had been forced to live for seventeen years as a white savage with his soul in a state of suspended slumber.[48]

Appointed by the government to a lonely position as lighthouse keeper, and having quarreled with his family, Pelletier was rumored to have grown morose and solitary, staring wistfully out to sea and flying into rages when villagers taunted him with the nickname *"le sauvage."* Around 1883 he married Louise Mabileau, a young local seamstress, but the couple had no children and few friends. Narcisse Pelletier died of unknown causes on September 28, 1894, at the age of fifty. Villagers speculated that he'd succumbed either to the long-term effects of Aboriginal sorcery or to a primitive nostalgia for his clan and country. Both explanations suggest that he was viewed as a failed Frenchman, unable to escape the legacies of "Endeavour Land" savagery.

Neither of the publications about the two castaways attained a wide circulation, even in their home countries. The pamphlet on Morrill was considered unsatisfyingly brief, and the expensive volume on Pelletier, not widely read in France, wasn't translated into English until 2009.

By contrast, Louis de Rougemont's fantastic plagiarism of their stories managed to reach popular audiences in half a dozen countries. His crude, if sometimes romanticized, inventions of Australian Aboriginal life and values for *Wide World Magazine* achieved a record circulation in Britain. The subsequent book is said to have sold more than fifty thousand copies, and was published in American and foreign-language editions. Even Australia's first acknowledged literary genius, Henry Lawson, who might have known better, was entranced by the preposterous memoir. He thought de Rougemont's style and sentiments "delightful," and claimed that the con man had made "a bigger splash in three months than any other Australian writer has begun to make in a hundred years."[49]

It was ever thus.

7

REFUGE

William Kent Escapes His Past

BORN ONE YEAR AFTER Narcisse Pelletier, in 1845, Australia's first professional Reef scientist experienced a childhood more bizarre than even the wildest inventions of a de Rougemont. William Kent's tortured path to the Great Barrier Reef can be said to have begun with an early-morning event in July 1856, at the age of eleven. That was the morning William and his sister Constance, twelve, ran away from their home in the small rural village of Road on the Wiltshire and Somerset borders in southwest England.

They'd planned their escape with care. The instigator, Constance, having retrieved some of her brother's old clothes that she'd secreted in a hedge, led the way to an unused garden privy screened by shrubbery. Here William helped cut off her long hair and toss it down the privy's drop. Now dressed as two little brothers, and armed with eighteen pence and a small stick, they set off for the distant port of Bristol. From there they intended to sail as cabin boys to the West Indies, and to somehow find their eldest brother, Edward, a junior officer on an inter-island steamship.

By evening they'd reached Bath, ten miles away. There the publican of the Greyhound Hotel, suspecting they were runaways, alerted the police. Under questioning, William broke down in tears and was returned to the hotel for the night. Constance, pugnacious and defiant, remained under lock and key at the station to prevent further escape. The following morning they were collected by a servant and returned home.

A Bath newspaper reported the event coyly as "A Little Romance," delighting in the brave behavior of the girl and the childish fancy of the escapade. As keen readers of exploration and adventure stories, the two children had often been caught playing daring games of make-believe on the roof of their substantial home. But the reporter missed the underlying desperation in their attempt to run away. Brother and sister returned to a harsh interrogation from their father, Samuel, and their hated stepmother, Mary Drewe Kent, née Pratt. William sobbed for forgiveness, but Constance remained unmoved, explaining simply: "I wished to be independent."[1]

Had the two children been able to articulate what really drove them, they would have told a pitiable story. Charles Dickens, with his nose for sentimental melodrama, adapted the incident in his last and unfinished novel, *The Mystery of Edwin Drood* (1870). Here he has the William figure, Neville Landless, explain his motivations for running away: "I have had . . . from my earliest remembrance, to suppress a deadly and bitter hatred. This has made me secret and revengeful. I have been always tyrannically held down by the strong hand. This has driven me, in my weakness, to the resource of being false and mean. I have been stinted of education, liberty, money, dress, the very necessaries of life, the commonest pleasures of childhood, the commonest possessions of youth. This has caused me to be utterly wanting in I don't know what emotions, or remembrances, or good instincts . . ."[2]

It was a plausible reconstruction of William's burning anger, but if anything, Dickens understated the misery of his two real-life models. Throughout their childhood their father, a sub-inspector of factories, had been gripped by the twin obsessions of social aspiration and sexual conquest, impulses that led to frequent moves and fractured, insecure lives for his large brood of children. William and Constance were shunted back and forth between a succession of unaffordable houses and austere boarding schools.

Samuel Kent's bullying personality and rampant libido had also broken the health, mind, and spirit of his cultured first wife, Mary Ann Kent. She was subjected to a stream of annual pregnancies, a syphilis infection that left both William and Constance with permanent legacies, and a regime of crushing isolation and humiliation as her husband conducted a flagrant affair with their live-in governess, Mary Drewe Pratt.[3]

Young William, timid and artistic, had clung to the company of his mother and his two elder sisters, while Constance, as tough and willful as her father, turned to furious rebellion as she neared her teens. By the time of their attempted flight to Bristol, the children's mother had died. Mary Pratt, now their triumphant stepmother, was already pregnant with the second of many half brothers and sisters. William was ridiculed by his stepmother as a girlish mommy's boy and made to use the backstairs of the house, like a servant. He was also forced to wheel his baby half sister through the streets of the village in a baby carriage, running the gauntlet of jeering local kids who loathed the dandies at Road House.

The ultimate consequences of Samuel's behavior proved too grim even for Dickens's later co-option. On June 30, 1860, two villagers discovered the body of William and Constance's three-year-old half brother, Francis Savill Kent, in the same outdoor privy where four years earlier Constance had changed into her runaway's clothes. Francis had been abducted from the house while he slept, smothered to death, and stabbed in the chest with a razor. His head was almost severed from his neck. Circumstances pointed to more than one perpetrator, and to them being from inside the household.[4]

Some thirty years on, William and Constance Kent made a second attempt to flee the shores of England. William, then aged thirty-nine, arrived in Tasmania in July 1884 accompanied by his wife, Mary Ann, and his half sister Mary Amelia. Late the following year Constance Kent, forty-two, left England on the *Carisbrook Castle*, reaching Sydney on February 27, 1886.

The once inseparable brother and sister had traveled under starkly different circumstances. The former, now a scientist, sailed in comfort to Hobart on the steamship *John Elder* to take up a position as superintendent of Tasmanian Fisheries. Constance sailed in steerage under the alias of Ruth Kaye, alone and penniless. She'd just completed twenty-two years of grueling imprisonment for the murder of her baby half brother.

Whether William had invited Constance to follow him to Australia or was hoping to avoid her by leaving England just before her release remains unknown. He had, however, written to her in prison in 1883, likely disclosing his intention to move to Australia with their four remaining half siblings. With or without his encouragement, Constance, as soon as she was freed in 1885, set sail for the distant convict continent. It was, as Dickens had so often shown, an ideal place to escape past shame and to fashion a new identity.[5]

William, like it or not, could never forget the debt he'd incurred to his sister. He owed Constance everything: his freedom, his reputation, his career, his comfort. The shrewd Scotland Yard detective who had led the murder inquiry, Inspector Jonathan Whicher, had been unable to find enough evidence to make a charge against the boy stand, but he remained convinced that Constance had carried out the murder in collaboration with William. Their motive, Whicher believed, was to exact revenge on their stepmother for blighting both their own and their mother's lives— an assessment backed by the most authoritative modern analysts of this famous Victorian crime.[6]

In the summer of 1865, five years after the death of Francis, Constance confirmed at least half of Inspector Whicher's suspicions by confessing to the murder. A bout of intense religious indoctrination had aroused her conscience, and she seems to have been moved by a martyr's determination "to absolve her family, especially William." Yet her overinsistence on having done the deed alone rang false with many of her interviewers. Inconsistencies and evasions in her account suggested she was protecting someone in the family, and she compounded this impression by refusing to take her counsel's advice to plead a family history of mental instability. To do so, she said, would damage her brother's future career. And so she stuck to her confession, even though she risked execution as a result—a sentence she did in fact incur, but which was later commuted to twenty years' imprisonment.[7]

Constance had always encouraged William's ambition to be a scientist. Samuel Kent and other family members wanted the boy to go into business in his maternal grandfather's coach-making company, but William's heart lay elsewhere. He was talented at drawing and he loved nature rambles, the seaside, and stories of science, exploration, and adventure. In 1859 Constance introduced him to the most important scientific work of the century. She read, and probably brought home, a copy of Charles Darwin's newly published *On the Origin of Species*, and then announced to her horrified parents that she was a convert to evolution. At the time of her sentencing in 1865, she knew, too, that William was shortly due to inherit money from his mother's estate, which could fund his education. A taint of family insanity might cast this legacy into doubt. Constance had thus performed an act of martyrdom to protect her beloved younger brother.[8]

In opting for a scientific career, William had not chosen an easy path. Clergymen steeped in the creationist tenets of the Church of England still monopolized most paid scientific positions in England at that time. Luckily, though, William's desire for a science education coincided with a social revolution that was beginning to pave the way for a new brand of professionalism in the field. Inspired by Charles Darwin's ideas and led by his pugnacious disciple Thomas Huxley, scores of young men from the insecure lower-middle class began to storm the traditional bastions of science and engineering, in the expectation of making a living. Huxley urged them to overthrow clerical and aristocratic privilege and bring about a "New Reformation," by forging useful, secular, and morally uplifting scientific careers.[9]

With the help of his recent inheritance William Kent was well placed to join these aspirants, even though his earlier education had been spotty and erratic. After his stepmother died in 1866 and his father moved to rural Wales to escape the murder scandal, timid William began to reveal abilities and ambitions few had suspected. Enrolling in evening classes at King's College, London, he attended Huxley's inspirational lectures on marine biology and decided to follow in his footsteps. Probably at his teacher's urging, he joined the Microscopical Society and began investigating minute marine organisms called infusoria. After this he worked for a period with Huxley's great friend William Flower at the Royal College of Surgeons, cataloging coral collections and becoming "smitten" with these mysterious creatures.[10]

By 1870 William, aged twenty-five, was a published authority on corals and sponges, and was working as a junior assistant at the British Museum. But he shared his father's restless ambition and chafed at the poor pay and lack of promotion opportunities. Three years on, the added financial burden of marriage to a London barrister's daughter, Elizabeth Bennett, goaded him into taking a better paid position as resident biologist at a new commercial aquarium in Brighton. It was an inspired decision for someone with both artistic and technological talents. William used the profits from the entertainment attractions of the aquarium to subsidize his scientific work—culturing lobsters and oysters, studying marine behavior and reproduction, and designing artificial marine environments.[11]

He worked at a succession of aquariums around the country, until the death of his first wife and his prompt remarriage in 1876 to wealthy Mary Ann Livesey opened the way for another promising career move. With financial backing from his new wife and support from the eminent zoologist Sir Richard Owen, William acquired a site on the island of Jersey in 1877, on which he intended to build a marine research station operating on the profits of an associated zoo, aquarium, and museum.

When this scheme collapsed through lack of government support, William decided to rejoin Thomas Huxley's patronage train. He enrolled in the professor's comparative anatomy course at Imperial College, dedicated the three-volume *A Manual of the Infusoria* to him, and in 1880 obtained a position working with the great man in the government's Fisheries Department. A year later William tried again to set up an essentially government-funded marine research station, but this, too, failed when another of Huxley's protégés was preferred. As a consolation, Huxley recommended William for a lesser but perhaps rather timely—given Constance's imminent release—position in Tasmania, supervising the state's fisheries.[12]

Whatever his other motives, the decision to go to Australia was partly a result of William's inability to obtain a scientific position in Britain to match his aspirations. Since his patron Huxley was by then the leading titan of British science, William had expected more. His was probably a failure of personality rather than talent. He'd always struggled to make close male friends, partly because he had inherited something of his father's tactlessness and arrogance, along with his obsession with money, status, and social advancement. William was prickly about his standing as "a gentleman," a status that was uncertain for such self-made men. As a result he was quick to take umbrage, and to engage in public disputes whenever he felt his honor was being impugned. Thomas Huxley approved of such ambition in his disciples, but he expected it to be laced with a dash of social idealism that seemed absent in William. In short, William was respected but not liked. Having inherited sterility thanks to his father's syphilis, he also lacked the children so important to life in Huxley's inner circle. All in all, William Kent's personality seemed repellently cold and calculating.[13]

These traits were revealed in his shameful treatment of Constance during her twenty-two years of miserable imprisonment. In all this time William wrote to her just twice, and then only for formal reasons. He also completely ignored her many pleas to visit. Throughout this long neglect, Constance battled with prison authorities over her harsh conditions and treatment, sending nearly forty petitions for early release, all of which were rejected. Though contrite about her crime, she felt an understandable bitterness at the extent of her sacrifice. As the completion of her sentence drew nearer, William thus had every reason to fear what his sister might say or do after her release. While she'd been serving her debt to society, he'd been chasing his career. If forced by circumstances, Constance could well decide she had little to lose by revealing the full truth of the murder.[14]

If William Saville-Kent, as he'd begun calling himself from around 1880, hoped that moving to Tasmania would fulfill his ambitions and eclipse the weight of the past, he was wrong. Feted by the government on arrival, he soon found the pay inadequate and the role unclear. This ambiguity entangled him in a vicious local squabble. The Salmon Commissioners, a powerful political lobby group with gentry pretensions, believed that his chief function was to confirm the acclimatization in Tasmania of British salmon. On investigating the evidence, however, William decided that the vaunted trophies of eminent figures like the governor weren't examples of the royal sporting fish, *Salmo salar*, at all, but overgrown specimens of European and American trout. Genuine salmon had failed to acclimatize because of the warm local waters. William's abrasiveness in publicizing this embarrassing opinion would, among other factors, eventually cost him his job.

As a disciple of Huxley, Saville-Kent had arrived in the colony with a scientific program that quickly proved too visionary for the faux-gentlemen of the Salmon Commission. After a systematic review of local fishery conditions, he immediately set about developing a research station to study and culture marine species. Built to his design on government land at Hobart's Battery Point, it boasted a laboratory, saltwater hatcheries, and aquaculture facilities. Here he instituted methods for culturing oysters and lobsters, and encouraged local fishermen to develop markets for indigenous fish and crayfish. In his spare time he wrote zoological papers, expounding to Tasmania's Royal Society his ideas about the state's failure to produce genuine salmon, and fueling his enemies' complaints that he was too academic for the superintendent's job.

By late 1887, William's situation in Australia looked untenable. An application to renew his Tasmanian contract was rejected, some part-time oyster-protection work he'd been doing with the Victorian government was due to terminate, and an overture for a fisheries position with the New South Wales government was blocked by a rival.[15]

William's other pressing mission, to acclimatize the scandal-haunted Kent family in Australia, had also hit some difficulties. True, his younger half sisters were doing well. Florence had arranged a position as a governess in Sydney before migrating, and was now trying to find a similar post for Mary Amelia. (Eveline later trumped them all, arriving in Australia in October 1889 as Mrs. Johnson, having married a Melbourne doctor in Europe.) But William's younger half brother Acland, twenty-six, had pulmonary tuberculosis which hampered his efforts to find work. After spending six months in Hobart, he'd drifted to the Victorian goldfields, where he died in Bendigo in May 1887.[16]

William traveled alone to Melbourne to attend the funeral of this last surviving brother, who'd been born only one month after the death of Francis. Twenty-seven years earlier, at the age of fifteen, William had sat beside Francis's tiny coffin on the way to the family graveyard at East Coulston, where a menacing headstone read: "Cruelly Murdered at Road/ June 30th 1860/Aged 3 Years and 10 Months/Shall not God Search this Out/For He Knoweth the Secrets of the Heart?"[17]

At this moment in 1887, too, Constance must have loomed like a ticking bomb, being now virtually on his doorstep. So infamous was she back in Britain that her deeds had been sung in ballads through London's streets, and her wax effigy displayed for two decades at Madame Tussaud's. Her release from prison had been reported in metropolitan newspapers. Yet what William did to contact or help her after she arrived in Sydney, nobody knows; nothing certain has been established of Constance's movements during her first three or four years in Australia. Perhaps William offered her money to support herself in Sydney, or perhaps he invited her to live in his home, first in Hobart and later in Brisbane, until she found work. Perhaps it was the strain of her presence that made Mary Ann decide to go back alone to England in October 1887, in order to take a "holiday."[18]

In the year of Australia's centenary, 1888, William's luck suddenly changed. An unsolicited request from the Queensland government to report on the state's main oyster sites in Moreton Bay was followed by an invitation to accompany Captain H. P. Foley Vereker of the HMS *Myrmidon* on a survey voyage of the Cambridge Gulf in northwest Australia. "With alacrity," William grabbed at this chance to escape his troubles and explore a tropical region unknown to science.[19]

Even the transit voyage on the China Navigation Company's steamer *Tsinan* proved life-changing. On the way to join the expedition in Darwin, the ship called for a few hours at the Cairncross Islets, around forty miles north of modern Mackay. William, who'd never seen a living coral reef, arrived at dusk, when the upper platform of the fringing reef was partially uncovered by the tide, to reveal corals "growing in their native seas and in their wonderful living tints."[20]

William threw himself into a frenzy of collecting, theorizing, and drawing. He sketched black bêche-de-mer (*Holothuria*) pushing particles of sand and coral into their circular mouths, purple starfish thrusting their spinous arms "in every direction apparently seeking for food," and semitransparent pink *Synapta* floating ethereally in shallow tidal rock pools. "Unsolved mysteries" seemed to confront him wherever he looked: the unknown taxonomies and ecologies of the thousands of chambered spiral shells scattered on the reef, clusters of young stony *Madrepore* corals floating on chunks of pumice, hordes of tiny oysters clinging to mangroves and apparently new to science. In short, William Saville-Kent fell instantly and permanently in love with the wondrous marine world of the Great Barrier Reef.[21]

A stint of collecting in Darwin before embarking on the longer survey of the Cambridge Gulf saw him gather further biological riches. These, on William's return, brought him introductions to several of Queensland's most influential scientists. He donated sixty-seven specimens of molluscs to the Queensland Museum, care of its biologist, Charles Hedley, and a substantial collection of birds and reptiles via the museum's zoology curator, Charles de Vis. Many proved new to science, including a bird that de Vis flatteringly named *Natricidiae kentii*. Reporting these discoveries in a paper to the Congress of the Australian Advancement of Science early in 1889 earned William an invitation to join the Queensland Royal Society, and, only a month later, nomination as the society's next chairman.[22]

Such triumphs added weight to a proposal for a full-time job that William had dashed off to the Premier of Queensland, Sir Thomas McIlwraith, just before joining the survey on the *Myrmidon*. His timing proved perfect: Queensland's two rival conservative leaders, McIlwraith and Sir Samuel Griffith, wanted to join political forces and both were worried about the depletion of the state's marine resources. All of the Reef's marine industries, including the lucrative pearl-shell trade, had been showing falling returns because of overfishing and the incursion of Japanese luggers. New licensing regulations passed in 1881 had not worked. By March 1889, even the Torres Strait pearlers were pressing for restrictions to be placed on the harvesting of immature pearl shell.[23]

In that same month, William received an offer from the Queensland government for a three-year, full-time position as Commissioner of North Australian Fisheries on a substantial salary. He immediately traveled to Brisbane by train, accompanied by his wife, Mary Ann, and another unnamed member of his family, probably Constance. The three moved into "Ellan Yannin," a comfortable Queenslander house high off the ground with wide verandas and trellises of climbing vines. Set in bushland at Kangaroo Point, overlooking the Brisbane River, it was soon filled with genteel objects, including a piano, stylish furniture, and a handmade shotgun for hunting birds.[24]

A well-paid job was not the only by-product of William's voyage north. While waiting for repairs to the ship in Darwin, he had gathered and recorded local fish species with the help of a policeman, Paul Foelsche, who was also a noted photographer. Mary Ann, on hearing William's subsequent account of the man's photographic achievements, seems to have been inspired to give her husband a "modest form of camera." This in turn prompted him to buy two baby fern owls from a local bird salesman to use as photographic subjects. Both actions proved portentous, the camera because it brought William's disparate artistic, scientific, and technological talents into a single unified focus, and the two baby owls because they liberated repressed feelings of love and whimsy in his wounded personality.[25]

William loved his camera and his birds, recording every antic, pos-
ture, and vocal acquisition of the two "balls of fluff" with the besotted
delight of a new father. As brother and sister, the owls displayed the same
bonds of affection that he and Constance had shown each other long
ago at Road House. Like the two children during that terrible time, the
owls played "with delightful abandon" in the presence of those they
loved, but shrank and froze into sticklike immobility in the threatening
presence of others. So deeply did William identify with the little creatures that he even visited the local zoo to chirrup greetings to its collection of downcast fern owls. He admitted also to a strong urge to "open
their prison doors" and let them fly free, a feeling his sister would have
understood only too well.[26]

If, as seems likely, Constance did share William and Mary Ann's
Brisbane home for part of 1889, it would have been a chance for the two
siblings to restore something of the broken trust between them. In the
winter of that year, Mary Ann and a female relative traveled down to
Sydney by rail on an unspecified errand. Soon after this, Constance's
movements can be tracked with some certainty. The year 1890 marked
the beginning of her new career in Australia. Starting as a volunteer in
Melbourne's typhoid tents, "Ruth Emilie Kaye," as she was now known,
went on to enroll at the Alfred Hospital, embarking on what would turn
out to be a long and distinguished nursing career.[27]

Released suddenly from the strain of unemployment and potential exposure as a murderer, William also discovered a refuge from worldly cares
and his past at the remote marine frontier of the Torres Strait, a place
where he could exercise his full range of talents and ease the shackles on
his stiff personality. After the setback in Tasmania, he was eager to show
off his practical usefulness and economic value as a marine scientist and
resource manager. A review of Queensland's fishing industries, plus conversations with McIlwraith and Griffith, suggested an urgent need for
what he called "a redemption" of the edible oyster and pearl-shell industries, especially the latter. The Torres Strait pearling industry, normally
one of Queensland's leading revenue producers at around $350,000 a
year, had become so exhausted that much of the harvested shell was now
too small for button manufacturers to use.[28]

Ever since his time as an aquarium biologist, William had championed artificial cultivation as a means of developing sustainable fishing industries. But nobody had yet come up with a way of doing this for pearl shell. Most of the region's shallow pearl-shell beds were exhausted, and deep-sea beds could not be protected from plunderers. Veteran pearlers also denied the viability of transferring immature oysters to shallow pools because this would mean severing their "abysses" (anchor cables), which would cause them to die or drift away on the currents. Moreover, luggermen believed that young oysters would perish in transit even before this.

Between 1889 and 1891 William made three extended stays in the Torres Strait to tackle these problems. Enthusiastic support from two of the region's most influential Europeans made his task a good deal easier. Frank Jardine was a pearler-adventurer from Somerset, near the Albany Passage, and John Douglas was Government Resident and Police Magistrate at Thursday Island: both helped William to visit the deepwater Old Fields near Badu Island, where, using boats and diving equipment, he was able to collect abundant samples of immature shell.

Jardine, his wife, Sania, and son, Bootles, also offered William hospitality in their cliff-top house at Somerset, and provided sites for the giant clamshells filled with seawater that he used for his oyster-transplantation experiments. Having successfully transported young pearl oysters from the deep beds in these portable aquariums, William also discovered that a series of shallow coral rock pools near John Douglas's Thursday Island residence offered current-washed environments perfectly suited to oyster growth. To his delight, the young pearl oysters "adapted themselves with alacrity to the novel environment." They grew new abysses until their shells were heavy enough to resist the currents by their own weight. In one six-week period most of William's young oysters added an astonishing half an inch to their shells.[29]

After canvassing opinions from both small- and large-scale pearlers, William recommended a program of industry reforms. In order to replenish oyster stock, he contended, Endeavour Strait needed to be closed to pearling for three years, and no pearl shell should be sold before reaching an interior measurement of six inches—a size enforceable by inspection. At the same time, accessible banks, foreshores, reefs, and shore stations around the strait could be leased to luggermen for the transfer and "cultivation" of immature shell. Such a measure would also reduce dangerous and expensive deep-sea diving, enable owners to stop their divers stealing pearls, and deter poaching by foreigners. In 1891 the

Pearl Shell and Bêche-de-Mer Fisheries Act of 1881 was amended to include all William's suggestions.[30]

Investigating a variety of edible oyster beds at the southern end of the Reef, around Wide Bay, Mackay, and Cooktown, also buoyed William's spirits. He became convinced, half wistfully, that there was no more "perfect elysium" than the life of a north Queensland oyster farmer. "In no other country in the world," he wrote, "is so healthy, congenial and non-laborious a means of earning a substantial competency open to . . . all classes." He helped these farmers combat the problem of losing a high percentage of free-swimming oyster embryos (spat) at sea by building cheap, split-paling collectors on which the spat could cling. He also urged the industry to develop advanced culturing processes whereby ova could be matured into viable embryos within hatcheries.[31]

Taken overall, William concluded, the twelve-hundred-mile extent of the Great Barrier Reef represented "a vast harvest-field ripe for the sickle, wherein, as yet skilled biological labour is all but unknown." Much could still be done, he suggested, to exploit lucrative and potentially sustainable marine resources such as edible fish, dugong oil, turtle meat, tortoiseshell, and black coral.[32]

Yet it was as a scientific rather than an economic biologist that William grew to love the Reef. He thought the region's intertidal areas and "lime saturated" coral seas to be "one of the most active and visibly effective of Nature's petrological laboratories." And Thursday Island, he believed, offered a site "unequaled" in the world for the study of tropical biology. To prove his point he threw himself into a breathless research program, drafting scores of scientific papers, collecting specimens for London's Natural History Museum, investigating new species with his dissecting knife and microscope, and sketching and photographing varieties of corals for a projected book on reefs.[33]

One intriguing mystery lay almost at his back door in Brisbane. While inspecting oysters at Moreton Bay—the huge saltwater bay into which the Brisbane River flows—he noticed substantial remnants of dead corals belonging to the prolific reef-growing genera *Madrepore* and *Favia*. But why had these once-thriving corals died? A change in climate bringing colder temperatures was one possibility. Or perhaps a geological elevation of the seabed had lifted the corals out of reach of the tides. He thought it most likely, however, that sandbanks had shifted to obstruct the inflow of seawater into the bay at the same time as it was inundated by river floodwaters. This combination would subject the corals to toxic doses of fresh water.[34]

So little was known at that time about the character and behavior of the different coral types within the Reef. The celebrated HMS *Challenger* expedition of 1872 had collected sixty-one species of Australian reef corals, yet William, who made no claim to exhaustiveness, discovered more than seventy different species of *Madrepore* alone. Different reef locations also gave rise to distinct combinations of coral types. *Madrepore* corals, like the staghorn, dominated at some Port Denison reefs; luxuriant, bush-like clumps of *Millepora* at some Palm Island ones; mushroom corals off Adolphus Island; and leathery, bright-green *Alcyonaria* corals at many of Thursday Island's reefs. Yet on reefs of the outer Barrier, such as Warrior Reef, he found no signs of the hummocky stony coral species then known as *goniastrae, meandrina,* and *astraea,* which were present on most inshore and fringing reefs.[35]

Reef corals also proved more protean than he'd expected. Colors often varied considerably among separate colonies of the same species, and even within what were clearly different growth epochs of the same colony. Though William did not say so explicitly, this seemed to imply environmental adaptation. Non-coral species, he noticed, also made a surprisingly large contribution to reef building. Scientists had long realized that other species with limestone structures were assimilated into coral reefs, because their skeletons were sometimes obvious in reef boulders. Now William gave an exhaustive list of the organisms that could be so absorbed. It included nullipore algae, whose tissues were lime-encrusted; minute protozoa from the class Foraminifera; sea urchins; starfish; and trepang, all of which appeared to contribute to the lime cement and conglomerate that made up the reefs.[36]

Knowing exactly how fast corals could grow suddenly assumed a dramatic significance, when on February 28, 1890, the pride of the British India and Australian Steam Navigation Company's fleet, the *Quetta,* hit a submerged coral pinnacle in a well-charted area between the Albany Passage and Adolphus Island. The huge steamship sank in three minutes, with the loss of around half of its 182 passengers and crew. Asked to investigate the site, William concluded that the coral pinnacle responsible had grown to become an unmarked hazard during the thirty years since the last survey. Accurate studies of coral growth rates in different environments needed to be undertaken urgently.[37]

When it came to studying the prolific Reef order of Actiniaria, which includes anemones, William received unexpected help. Soon after arriving in the Torres Strait he met Alfred Cort Haddon, a younger Cambridge scientist with a reputation for his work on British anemones. Worryingly at first, Haddon's mission on the Reef seemed identical to William's own: "I propose to investigate the fauna, structure and mode of formation of the coral reefs in Torres Straits . . . to map the raised and submerged coral formations, . . . to investigate the fauna of the lagoons of the shore exposed at low tide and of the submarine slope . . . to endeavour to determine the zones of different species of coral and of associated invertebrates, and also what conditions of light, temperature and currents are favourable or otherwise for the different species."[38]

As it turned out, the men did not in practice compete, because Haddon's growing fascination with Islander culture eventually took him to anthropology. In the meantime the two Englishmen were delighted to pool their knowledge. Each named a new species of anemone after the other. William's discovery, the giant twenty-four-inch *Discosoma haddoni*, had the additional attraction of a "commensal" relationship with a small colored fish and a pink-striped shrimp. He speculated that the fishy visitors paid for their safe haven by serving as lures to attract other marine creatures into the anemone's mouth.[39]

William believed, too, that he'd acquired an advantage over most marine biologists by learning to use photography as a tool for the scientific study of reefs and their inhabitants. In a pre-scuba world, living corals were rarely seen because they grew under water. Even photographing them was possible only during fleeting periods of exposure at low spring tides. Then, and provided one worked with great speed, corals could, William boasted, "be reproduced with the fidelity that photography alone can compass," and that no pencil could equal. Photography could reveal the geological structures of reefs, map the distribution and relationships of reef corals, and capture the exact likenesses of marine species while they were still brimming with life. Photographs could also serve as precise records of changes in coral growth and distribution over time.[40]

To achieve all this, a scientist-photographer needed great patience and physical stamina as well as technological virtuosity. Specialist equipment for scientific photography was nonexistent. William had to devise his own square lens frame, and he built an extra supporting leg on his tripod in order to take shots of corals and tiny crustaceans from a vertical position. He experimented endlessly with different lens types to find the most suitable focal lengths for capturing the true size of his specimens. Wherever possible he photographed marine creatures, other than living corals, in his portable giant-clamshell aquariums, taking care to replicate the original environmental conditions and retain true natural appearances.[41]

Capturing the physical exactitude of this marine world was not William's only mission. The adventure, beauty, and romance of this "fairy land of fact" also struck deep chords in his personality. Thursday Island, his headquarters in the Torres Strait, was still a wild frontier pearling port with a population of only two thousand, made up of peoples "from every quarter of the globe." The year before William arrived there, the Government Resident had listed twenty-four nationalities among the annual list of offenders in the jail book.[42]

One could not imagine, for example, a more swashbuckling character than William's friend Frank Jardine, who was now living for periods on Thursday Island. Jardine had once herded his father's cattle twelve hundred miles through unexplored Cape York bush, fighting Aboriginal warriors all the way, before reaching Somerset in a tattered emu-skin suit. There he'd married Sania Solia, a niece of the king of Samoa, and set himself up as a type of Reef baron. He was brutal in his suppression of local Aborigines, and liked to serve European guests their meals on silver plate made from coins looted from a nearby Spanish wreck.

The Jardines were just the type of friends needed to help finance a quixotic secret hobby that William had begun to develop while staying in the Torres Strait. He showed them the results of his experiments to introduce irritants into a living oyster so that it would create nacreous layers around them. By this means he'd created artificial "blister" pearls, which grew out of the pearl shell. But he also tantalized the Jardines "with hazy glimpses of a royal road to the rapid accumulation of untold wealth" by claiming to be on the way to achieving the holy grail of producing "freely detached" cultured pearls.[43]

Lean, bearded, and angular, wearing a solar hat and a trim-fitting suit, William cut a romantic South Seas figure and liked to photograph himself wading through lagoon shallows, camping on a beach in a grass hut, or working on his clam-shell aquariums. Thursday Island, with its reputation as a maritime badlands, suited William's boyish self-image of a dashing adventurer. It probably also offered chances for adventures of a more amorous kind.

On William's third visit there, in 1891, he formed an intimate friendship with the famous flower painter Ellis Rowan, when both were staying at the Grand Hotel. While she sketched flowers he strode "out on the rocks hunting for flowers of a different kind—sea blossoms." They talked, walked, fished, sketched, sailed, and stayed with the Jardines at Somerset. "I have rarely left a place with greater regret . . ." she wrote wistfully. Perhaps they simply shared a love of art—the basis of William's long friendship with elderly Tasmanian fish painter Louisa Ann Meredith.

Still, William did later, in 1894, create a scandal by running off to Melbourne with Louisa's young granddaughter, an action that led to

Saville-Kent's corals. Figure 3 (upper right) is an illustration of
Madrepora kenti. In *The Great Barrier Reef of Australia: Its
Products and Potentialities* by William Saville-Kent, 1893

calls for the "seducer" to be shot for his "dastardly crime." Perhaps these artistic women reminded William of his cloistered boyhood, when the conversation, painting, and poetry of his mother and older sisters had provided such solace. On the other hand, this may have been another instance where he took after his father.[44]

In any event there is no doubt that the land and seascapes of the Reef appealed powerfully to William's artistic sensibilities. He thought of his photographs as both scientific records and reefscapes, imbued with aesthetic beauties of color, design, and poetic evocation. Skull Reef, on the

A. FLOTSAM.—WRECK OF NEW GUINEA MISSION SCHOONER "HARRIER."

W. Saville-Kent, Photo.   London Stereoscopic Co. Rep.

B. JETSAM.—STORM-STRANDED CORAL-ROCKS, CAPRICORN ISLANDS REEF.

Flotsam and *Jetsam*, in *The Great Barrier Reef of Australia: Its Products and Potentialities*, 1893

outer Barrier, for example, reminded him—perhaps all too poignantly—of a decapitated human head with an "unevaporated tear in its eye." It was the aesthetic principle of sublimity, too, that drew him to produce a brilliant matching pair of photos labeled "Flotsam" and "Jetsam." One showed the stark, stranded hulk of the mission schooner *Harrier*, the other a series of colossal storm-stranded coral boulders.[45]

Mostly he sought to show "from an artistic viewpoint," using chromolithography to hand-color his drawings, the stunning visual patterns of the Reef's coral gardens and marine creatures. At Crescent Reef, also on the outer Barrier, he encountered:

> the most luxuriant expanse of living coral [he] had the good fortune to photograph . . . [I]n some examples . . . the corallum was bright violet throughout, with a tendency to magenta towards the tips of each separate branchlet; in others a creamy hue predominated, with violet or crimson extremities and growing points; while in a third series, the ground colour varied from light to dark sage-green, all the growing points . . . being violet or crimson.[46]

In December 1891, at the urging of his homesick wife, William left Queensland to return to England. In the British autumn of 1893 he published a book that encapsulated his four magical years of work and pleasure on the Reef. Published by the elite press W. H. Allen, it was a large-size production in super-royal quarto, measuring 13½ inches by 10 inches, with forty-eight full-page, photomezzo-type black-and-white plates, and sixteen hand-drawn, hand-colored chromolithographic plates.

The Great Barrier Reef of Australia: Its Products and Potentialities took its many reviewers in Britain and Australia by storm. They described it as "sumptuous," an "*edition de luxe*," with none complaining of the relatively expensive price of four guineas. They called it a unique kind of scientific work, one which covered its many themes in such multifaceted and compelling ways that every type of reader was satisfied. It was, we can now see, the first complete biography of the Reef. William Saville-Kent's wounded sensibilities, diverse talents, and frustrated ambitions had come together to produce a masterpiece.

At a time when Thomas Huxley and the poet Matthew Arnold were arguing about the emergence of a gulf between the two cultures of art and science, William had shown how to bridge this divide. The *West*

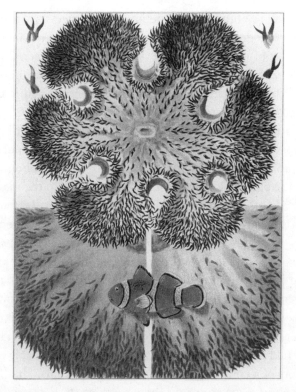

Giant anemone, named *Discosoma kenti* after William Saville-Kent.
In *The Great Barrier Reef of Australia: Its Products and
Potentialities*, 1893

Australian proclaimed the book was enough "to make the scientific man an artist and the artist a scientist, and to inspire the ordinary reader with a desire to be both." William's photographic illustrations, wrote *The Field*, showed "the beauties of the corals and other animals constituting these marvellous structures with a degree of accuracy which has never been even attempted."[47]

William had at last attained his twenty-five-year ambition to become a famous scientist. The *Saturday Review* echoed other reviewers in asserting that such a complete study of a coral reef had not been published before. *The Scotsman* declared it the most original book on coral reefs since Darwin's publications in 1836, and one destined to be always the "first authority on its subject." *Nature*, already on its way to becoming the most prestigious scientific journal of the Anglophone world, suggested that Saville-Kent's photographic methods had added something entirely new to the methodology of the scientist: "[his] book contains a series of nature-pictures of the corals such as has never before been submitted to the scientific world, and a glance at his illustrations does more to familiarise one with the phases and aspects of the reef and its life than pages of written description."

Australian journalists, and especially Queensland's leading newspaper, *The Courier*, hailed in particular the book's promotion of the Reef's economic products and potentialities. *The Argus* in Melbourne had no doubt that the publication of "such a magnum opus in the mother country" would advertise "the marvels of the Great Barrier Reef and . . . the magnitude and variety of resources . . . awaiting development." British newspapers like *The Times* and *Saturday Review* were especially impressed that William's lively writing style had managed to make the dismal science of economics read like "a veritable romance of the sea."[48]

And every reviewer, without exception, singled out the photographs and chromolithographs, hand-drawn and hand-colored, as the book's chief attraction, noting that most people in the Northern Hemisphere could have had no conception until now of the indescribable beauty and riotous colors of a coral reef and its marine inhabitants. "It almost takes our breath away to be suddenly shown one of these plates," wrote the *Cambridge Review*, "we feel we are looking at the thing itself, and we are lost in admiration at the skill of the photographer and the care of the publisher which have combined to produce these results." The *West Australian* thought William's artistry to be nothing short of genius: "Unless one has . . . seen for oneself the submarine chromatic effects which are more brilliant than the most gorgeous transformation scene conceived, it would be almost difficult to believe that the bright greens, reds, pinks, blues and yellows are the actual colour of forms . . . Scarce a flower upon earth can vie in brilliancy of tint with many of the anemones of the oceans, while the birds of the tropics find their plumage dulled besides the remarkable fishes which are found in these coasts."[49]

As many reviewers predicted, William's book made the Great Barrier Reef a place of celebration rather than notoriety, revealing its astonishing beauty and diversity to people with no idea of the existence of this tropical underwater world, and countering the negative perceptions arising from the stories of Eliza Fraser, Curtis, and de Rougemont. William's photography also helped make coral biology an intriguing subject,

Saville-Kent captures the color and multiplicity of Barrier Reef
fish in *The Great Barrier Reef of Australia: Its Products
and Potentialities*, 1893.

and when Maurice Yonge and his Cambridge expedition went to the
Reef (the story of which is to come in chapter 10), Saville-Kent's was the
only scientific book about the Reef they'd ever seen.

While completing the book, William had been offered a three-year contract by the West Australian government to work as a Commissioner of
Fisheries. Leaving homesick-prone Mary Ann behind in England, he rep-
licated his scientific and economic successes there, including publishing a
pioneering study of the ecology of the Abrolhos Reef system, to the west
of Geraldton. That Constance was working in Perth as a nursing sister for
part of this time was possibly an additional attraction. On returning to
Britain in 1895, William produced a similarly sumptuous record of his
West Australian experience called *The Naturalist in Australia* (1897).

Great Barrier Reef Fishes, in *The Great Barrier Reef of Australia: Its
Products and Potentialities*, 1893

Settling down to retire in Britain wasn't easy after those years of tropical adventure and glamour. He and Mary Ann moved through a succession of houses in Chiswick, Croydon, and coastal Hampshire. For a time
William also haunted the musty clubs of retired Anglo-colonial gentlemen, and lectured to amateur buffs at the Bournemouth Natural
Science Society. But this all felt too superannuated for someone still har-
boring unfulfilled dreams and a fierce nostalgia. "Is [it] to be wondered,"
he asked in the conclusion to his *Naturalist in Australia*, "that emigrants
of . . . but a few year's standing only . . . [in] Australia's prolific soil and
sunny clime, find it difficult to rehabilitate themselves contentedly amidst
the grudgingly responsive fallows, predominating fogs and murky skies
of their native land?"[50]

His hints in the pages of *The Great Barrier Reef* about the potential
riches of pearl cultivation eventually found some takers. In 1904 William
returned to Thursday Island in the employ of the Lever Pacific Plan-

tations Company, to transport fifteen hundred pearl oysters to the Cook Islands for cultivation. Although this experiment failed, his secret work on artificial pearls apparently made better progress. Returning to Queensland in 1906, he formed a pearl-culturing company with British and Australian financial backing, and leased a section of the Albany Passage adjacent to the Jardines. Bootles, who assisted with the new experiments, would later claim that William did actually succeed in culturing freestanding artificial pearls, but if he did he never reaped the rewards, which went to Japanese rivals.[51]

William Saville-Kent's unpatented pearl-cultivation method was one of two great secrets he carried to his grave when he died suddenly, on October 11, 1908, near Bournemouth, of a blocked bowel. Mary Ann sold all his books, papers, collections, and menagerie the following year. When read by others, his notes on culturing artificial pearls proved unintelligible.

Still, someone cared enough to decorate William's gravestone with a symbol that he would surely have valued above any other—a collection of Great Barrier Reef corals. It would be nice to think that Constance put them there, but she never returned to Britain. As Ruth Emilie Kaye, she became an esteemed nursing sister and matron, and in her hundredth year received a letter of congratulation from the Queen, before dying on April 10, 1944 in Strathfield, Sydney.

Having worked in typhoid tents, a leper hospital, and a reformatory school for girls, Constance can surely be said to have redeemed herself. If the Reef could speak, perhaps it would say the same of her brother William.

8

PARADISE

Ted Banfield's Island Retreat

I N 1908 READERS IN THE WESTERN WORLD were introduced to another publishing sensation from Australia, in the form of a new version of an old myth. *The Confessions of a Beachcomber*, written by E. J. Banfield, told the story of two modern-day Robinson Crusoes who'd abandoned civilization to live on a small tropical island within the Great Barrier Reef lagoon. Dunk Island, named by Captain Cook after an Admiralty dignitary of the day, was one of the Family group of islands and islets around 110 miles north of Townsville and two miles from the mainland coast.[1]

Edmund James (Ted) Banfield, a Townsville journalist of forty-four, and his wife, Bertha, a music teacher aged thirty-six, first visited the deserted island of three and a half square miles in mid-September 1896, while hunting for a site on which to build a getaway cabin. Like many Reef islands close to shore, Dunk had once been part of the mainland, and it was mountainous, wooded, and picturesque. Though it lacked the swaying coconut palms emblematic of the South Seas, it featured a fringing coral reef, a white sandy beach, and tall cliffs covered in trees and plants. A few hundred yards inland grew a forest of vine-entangled bloodwoods, Moreton Bay ash, swamp mahoganies, "Gin-gees," and native figs. Varieties of acacia, pandanus, and flowering hibiscus shrubs edged the strand, and green webs of native cabbage scrabbled down the beach.[2]

No sooner had the Banfields landed on the crescent bay at the northern end of the island than a canoe appeared, paddled by an Aboriginal man called Tom, one of Dunk's few living original inhabitants. Tom, who had been born on the island and belonged to the now scattered and fragmented Bandjin and Djiru clans, had somehow learned of the Banfields' intended visit and come over from the mainland with his mother-in-law, wife Nellie, and their nursing infant. To this tall, burly man with ribbons of scarification across his chest, this island was Coonanglebah, his lifelong estate, clan hunting ground, and Dreaming place. He knew its legends and habitat in microscopic detail, though he couldn't lay claim to a square inch of it, and he conducted Ted proudly

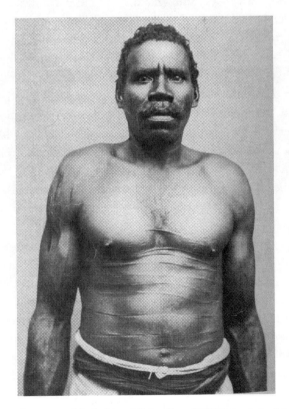

Tom of Coonanglebah (Dunk Island), c. 1898 (John Oxley Library, State Library of Queensland)

on a tour of the island's main attractions, including its priceless fresh-water streams.

Entranced, Ted stood on a long, sheltered plateau above the strand and ritually fired a rifle bullet into a bloodwood tree to mark the spot where they would one day build a house. Clouds of colored butterflies hovered over the beach, inspiring him to name the bay Brammo, after the poetic word for butterfly used by Palm Island Aborigines. Overhead, the trees shivered with colonies of white nutmeg pigeons and metallic starlings. This single visit was enough, Ted later claimed, to force an immediate and "revolutionary change" in the couple's outlook and future plans.[3]

Almost exactly a year later they returned to the island for a six-month trial stay, which, though it extended into years, was initially less than idyllic. During the intervening months they'd acquired a lease from the Queensland government for 128 acres of the best land, but Ted's mental and physical health had disintegrated. Small and slender at the best of times, his weight had plunged to an alarming 116 pounds and he needed chloroform to sleep at night. Diagnosed with the wasting disease phthisis by his Townsville doctor, he'd been given between three and six months to live. This stark sentence precipitated his and Bertha's decision to retreat to the island.[4]

When the Banfields landed at Brammo Bay on September 28, 1897, their new servant Tom had to carry Ted from the boat to a blanket that had been specially laid down above the tideline. "Ready to faint from weariness and sickness," he lay there longing to be home among the comforts of Townsville, while Tom and a hired workman lugged all the provisions, tools, and materials up to the plateau where the couple hoped to build a hut.

The following morning, Ted later claimed, he awoke to "a perfect combination of invigorating elements. The cloudless sky, the clean air, the shining sea, the green folded slopes of Tam o' Shanter point opposite, the cleanliness of the sand, the sweet odors of the eucalypts and the dew-laden grass, the luminous purple of the islands to the southeast; the range of mountains to the west and northwest, and our own fair tract—awaiting and inviting . . . Physic was never so eagerly swallowed nor wrought a speedier or surer cure."[5]

Despite this miraculous regeneration, forging a new way of life took a little longer. Ted and Bertha initially slept in tents and ate Tom's daily supplies of fresh fish, pigeon, and scrub fowl under the nearby bloodwood tree. Though still "in a frail physical state," Ted helped his two workers clear an area of scrub, bolt together the cedar home, and begin work on a kitchen and veranda extension.[6]

They felled a bloodwood and a bean tree, inched the logs to the site with a crowbar, and sawed them into rough planks. For foundations they mixed local stone with sand and tar. They scrounged posts and ridgepoles from the jibboom and masts of two wrecked ships that had drifted up onto the beach. The roof was made from cheap corrugated iron, the floor surfaced with beaten clay. And Ted, despite his "blistered and bleeding hands, aching muscles, and stiff joints," molded an assortment of twisted jungle timbers into crazy but effective furniture. The overall result, he claimed, was a tropical counterpart of the log cabin that his guru, the American nature writer and philosopher Henry David Thoreau, had built on the shores of Walden Pond in Massachusetts around fifty years earlier.[7]

Ted boasted that the shack offered no violation to "the genius of the Isle." It was "a little shambling structure of rough slabs," deliberately "unobtrusive" and "hidden in a wilderness of leaves." He and Bertha shared the interior with a multitude of geckos, spiders, grubs, and swooping bats, who treated it as a type of cave: "the low walls, unaspiring roof, and sheltering veranda [are] so contrived as to create, not tickling, fidgety drafts but smooth currents . . . [that] flush each room so sweetly and softly that no perceptible difference between the air under the roof and of the forest is at any time perceptible."[8]

To match the hut, Ted began evolving a lifestyle like Thoreau's, albeit a tropical version. On an island in the Great Barrier Reef, he said, "the career of the Beachcomber" offered "the closest possible 'return to Nature.'" All year round Dunk provided "the tonic of the sea and the Majesty [of] the Sun," which made for one of the most benign and equable climates on the globe. Influenza and all the other debilitating physical and mental sicknesses of the city, he claimed, were unknown. The odd bout of malaria troubled the couple less than the common cold. Clothes were hardly needed in any season—Ted wore only shorts and a large hat to shade his beaky nose and cowboy-style mustache. Bronzed and barefooted, he soon acquired the lean, muscular physique of a sailor. His weight climbed to 142 pounds, and he found he could labor in the sun all morning and swim in the clear warm waters of Brammo Bay all afternoon.

He felt that his entire sensorium had been revitalized. Scents of flowers, shrubs, birds, and marine creatures beguiled his nose. His ears became attuned to "the hum of bees and beetles, the fluty plaint of a painted pigeon far in the gloom, the furtive scamper of scrub fowl among leaves made tender by decay, the splash of startled fish in the shadows." As a beachcomber, he'd cast off civilization's discontents.[9]

At the time when Ted was using it, the term "beachcomber" was generally a pejorative one. It had originated in, or at least become widely applied to, the South Seas, by writers like Robert Louis Stevenson, and it described feckless, opportunist, hobolike characters who foraged for goods washed up from shipwrecks and lived off the sale of them. Ted's use of the term was partly ironic, and partly a reshaping to denote an altogether different type of person, one who relied on the provisions of nature to live a simple but ethical, aesthetic, and sensual life.

Ted, like Thoreau, now saw the townsman's pursuits as "devoid of purpose, insipid, [and] dismally unsatisfactory." By shucking off "the poisonous years of the past" and the "artificial emotions of the town," he and Bertha had attained a genuine "independence." Unfettered by mortgages, they could live comfortably on around $250 a year. No gourmet's feast surpassed the pleasure of eating fresh fish followed by homegrown pawpaws and golden mangoes. Ditching all schedules, they could "dally luxuriously with time" and "loll in the shade of scented trees, or thread the sunless mazes of the jungle . . . or bask on the sand." "The Beachcomber . . . is an individual whose wants are few—who is content, who has no treasure to guard, whose rights there is none to dispute; who is his own magistrate, postman, architect, carpenter, painter, boat-builder, boatman, tinker, goatherd, gardener, woodcutter, water-carrier and general labourer."[10]

Ted felt he exploited nobody. His daily bounty was thrown up on the sand, a tantalizing lottery generated by the chance actions of tides and currents. Today it might be a cedar log, tomorrow a weathered ship's figurehead; someday, perhaps, the prize of a black or pink pearl lying inside the flesh of a goldlip or blacklip oyster. Living on an "Isle of Dreams," the beachcomber was rich beyond imagination:

All is lovable—from crescentric sandpit—coaxing and consenting to the virile moods of the sea, harmonious with wind-shaken casuarinas, tinkling with the cries of excitable tern—to the stolid gray walls and blocks of granite which have for unrecorded centuries shouldered off the white surges of the Pacific. The flounces of mangroves, the sparse, grassy epaulettes on the shoulders of the hills, the fragrant forest, the dim jungle, the piled up rocks, the caves where the rare swiftlet hatches out her young in gloom and silence . . . all are mine to gloat over.[11]

He'd gained humankind's most precious state: "freedom—freedom beyond the dreams of most men in its comprehensiveness and exactitude."[12]

The Banfields lived in their simple shack from 1897 to 1903, but the realities of their life during this period bore little resemblance to Ted's later depiction of it. He'd fled to Dunk Island less as a rebel against commerce than as one of its failures, in the hope that the place might offer a fresh stab at business success. A child of the Australian frontier, Ted longed to match his father's pioneering achievements by helping to build a new commercial civilization on the Reef.

Jabez Walter Banfield, a sober, God-fearing Liverpool printer, had migrated to Australia in October 1852 to try his luck panning for gold along the Ovens River in northeast Victoria. Providing services for booming gold towns proved a better bet than chasing alluvial seams, so he set up a newspaper in the Victorian gold camp of Ararat and summoned his family in England to join him. After that, he'd risen with the town to become its leading burgher—a press proprietor, Justice of the Peace, magistrate, asylum patron, and celebrated local thespian.[13]

Ted, born in 1852, didn't see his father until the age of two, but he instantly worshipped this domineering pressman. The boy strained to emulate his father's talents and tastes, but with little success. Ted, small and intense with a puny frame, had a slightly palsied hand, and a watering right eye from a bicycle accident. All this confirmed Jabez's opinion that his son was "something of a lame duck." Moreover, he seemed neurotic and impractical, a talker and dreamer. By contrast, Jabez identified with his capable, self-confident eldest son, Harry, whom he anointed to take over the newspaper. Overshadowed and often miserable, Ted blamed himself for his father's contempt, seeking escape and consolation in solitary bush walks and the nature philosophy of Thoreau.[14]

Determined to prove himself in his father's trade, he moved in 1882 at the age of thirty to the Reef port of Townsville, where he helped an entrepreneur, Dodd Clarke, start a newspaper. Over the next decade Ted's decision seemed vindicated, as he chalked up a string of civic and business triumphs in the vein of his father. He'd arrived in Townsville on the crest of a boom; the port was a service center for the northern goldfields and also the site of an expanding sugar industry. A trio of ambitious local businessmen was pushing to establish Townsville as the capital of a separate northern state, free from the interference of Brisbane. Robert Philp, a shipping and retail magnate; Thomas Hollis Hopkins, a merchant; and Thankful Willmett, a publisher, were quick to recruit the impressionable new journalist to their cause.[15]

With their help, Ted was able to fund a trip to England in 1884 to consult a specialist about his deteriorating vision. The subsequent trauma of having to lose his infected right eye in an operation was offset by meeting Bertha Golding, the daughter of Liverpool family friends. Bertha's own affliction of partial deafness gave the couple an instant bond, but Ted also fell in love with her musical talent, good humor, and sharp common sense. They were married in Townsville in August 1886, and she won Jabez's approval when the couple eventually visited Ararat. Flushed with hubris after taking over the editorship of the *Townsville Daily Bulletin* from his sick boss in 1889, Ted started writing boastful letters to his father, airily offering him advice on mining investments provided by magnate friends.[16]

The following year saw Ted's bravura crumble. Dodd Clarke suddenly decided to resume his editorial position, relegating Ted to a humiliating downgrading. Further loss of face followed when he had to admit to Jabez that he'd lost his and Bertha's combined savings on his mining speculations, and this at a time when his wife's worsening deafness meant she had to stop giving music lessons. Ted's depression deepened as the north Australia separatist cause fragmented in the face of opposition by British investors, as well as rivalry from the nearby sugar port of Mackay, and white trade-union hostility to the use of indentured Islander labor in the sugar industry. Infuriated, Ted published proseparation harangues in the paper that were so extreme Dodd Clarke eventually had to intervene. Friends grew worried at Ted's emotional brittleness.[17]

Ted would later claim that his breakdown and flight to Dunk Island in 1897 was fueled by a hatred of commercial civilization, but this wasn't what he told family and friends at the time. His letters to Jabez and Harry represented the move as a shrewd investment in low-cost virgin land, ideally suited to the growing and sale of tropical fruits and other fresh produce. Cheap black labor, fertile soil, and a climate of high rainfall and steady sun would underpin production, while the weekly steamer visits to the island would enable distribution to markets on the nearby mainland.[18]

The plan proved easier to explain than to execute. For a start, the couple's early life on Dunk was not as easy or healthy as Ted would later make out. His daily diary entries from January 7, 1898, show that the onset of the rainy season brought him and Bertha repeated and disabling doses of malaria and dengue fever. Bertha was additionally prostrated with bouts of internal pain that only ceased after she underwent an operation and three months' recuperation in the Townsville hospital. There were also money worries. Before launching their proposed enterprise, they'd had to convert their landholding of 128 acres to freehold. This had to be financed. So, after only two years on the island, Ted resumed writing paid pieces for local newspapers.[19]

More than anything, Ted's diaries reveal the couple's dependence on the labor and skills of a succession of male and female Aboriginal workers: Tom, Nellie, Jinny, Mickie, Jenny, Toby, Sambo, Willie, Charlie, and others. Almost every day Tom or Mickie delivered fresh food from the sea, harpooning rockfish, shark, dugong, turtle, parrot fish, and much more. Ted seems to have been an indifferent fisherman himself and a worse sailor. Mostly he delegated the skippering of his boat to Tom, Toby, or Sambo, who picked up weekly supplies from the steamer and collected urgent items from the mainland ports of Cardwell, Bicton, Geraldton, and Townsville. While attempting to show off his sailing skills to Bertha in September 1899, Ted managed to capsize their boat in a sudden gust. Unable to swim, she was too traumatized ever to sail with him again. While sailing solo later that year he capsized once more, making so lethargic an attempt to avoid drowning that Bertha worried that he nursed a latent death wish.[20]

Most of their Aboriginal workers' tasks were physically grueling. They cleared and fired dense scrub, hacked down jungle, hoed and planted an array of vegetables and tropical fruit seeds, and then weeded, harvested, packed, and transported the products for sale on the mainland. They erected fences against snakes and eagles, and built hen and duck houses; they collected fowl eggs, oysters, crabs, and crayfish to add to the island's exports, and bottled and sold the abundant supplies of honey generated by Ted's dozen hives of Italian bees. The daily duties of the "gins," as Ted called the women, were no lighter: Nellie, Jinny, and Jenny had to weed, hoe, collect shellfish, and chop firewood, as well as cook and clean. As Ted's energies grew, he also supervised a flurry of ambitious developments: the building of a boatshed at the back of Brammo Bay beach and a suspension bridge over the gully, the laying of timber rails for a boat trolley, the installation of a storage tank and pump, and the planting of dual lines of coconut trees leading up to their hut.[21]

Ted's sweeping formal avenue and his flagstaff for commemorating ceremonial occasions suggests he'd begun to think of himself as the viceroy of an island empire. And as with imperialists everywhere, this aspiration led to the commissioning of grander premises and an expansion of his territorial domain. In January 1900 he decided to replace the hut with a professionally built bungalow. Six months later, he and Bertha gained approval to extend their landholding to 320 acres. Some of the money for all this came from a small legacy received on Jabez's death in December 1899—news that caused Ted to grieve that he'd now never be able to justify his move to his father. Perhaps the grandiose, fin de siècle plans for Dunk Island were his attempt to silence this most insistent of ghosts.[22]

In any case, Jabez's legacy was too modest to finance all of Ted's ambitious plans, so he increased his newspaper work by writing regular leaders for the *Cairns Argus* and other papers. In February 1901 he confirmed his recidivism by agreeing to take over editorial duties on the *Townsville Evening Star* for six months. It seemed a portentous decision: Ted was resuming the kind of work that had caused his breakdown, and to do it he and Bertha were returning to civilization. Backsliding to their former life after only four years on Dunk Island, the couple appeared to have come full circle.

In the end they had a nine-month absence from Dunk, and when they returned in November 1901 it was to a crisis. Most of Ted's bees had perished, shrinking the honey supply to nothing—and honey was their most lucrative export. The culprits proved to be two species of island birds: the Australian bee-eater and the white-rumped wood swallow.

The honey or the birds: Which had to go? This dilemma tested Ted's deepest values, because his love of birds was fundamental. After wrestling with his conscience, he decided that a profitable honey business couldn't justify the slaughter of such lovely creatures, which were after all only following their natural instincts. Discussion of the issue with Bertha led to a full-scale "review" of their philosophy of life and to the articulation of a new "grand objective." They would give up honey production and other forms of commerce to turn Dunk into a bird sanctuary, where all but raptors would be protected—the hunting of the latter being a sentimental exception that Ted soon abandoned in practice.

Although it was the honey crisis that triggered this change, other forces had been pushing in the same direction. One was Ted's reaction to rejoining the urban rat race for nine months. Although Bertha had enjoyed her respite from mosquitoes and loneliness, she realized that being away from Dunk had crystalized Ted's love of island life. Somewhat surprisingly, though, returning to full-time journalism had helped him recover his muse. Belting out political leaders for the local newspaper was as tedious as ever, but he'd found new delight in writing Dunk Island nature pieces.

Once their house was finished in 1903, reading and writing began to occupy a growing chunk of Ted's daily routine. As well as returning to his boyhood love of the romantic naturalism of Gilbert White's *Selborne* and Thoreau's *Walden*, he began studying works of marine and ornithological science. These included *The Great Barrier Reef* by William Saville-Kent, which Ted praised highly. He also began corresponding with several Queensland scientists who shared his floral and zoological interests.[23]

Giving up the commercial production of foodstuffs meant that Ted could spend more time exploring the island and its surrounds, which he did under the guidance of Tom, Mickie, Willie, and others. Ted began this education with many of the typical frontier prejudices of his day. He thought that Aborigines were a doomed Stone Age race, a childlike people incapable of rational thought, discipline, or morality, who treated the island as a larder for mindless consumption. But gradually Tom and

Mickie, in particular, instilled in him a different kind of understanding and appreciation of Dunk's marine, plant, and bird life. They taught him Aboriginal maritime, fishing, and bush skills, and they revealed to him marvelous hidden places and histories.[24]

It was Tom who led him to each of what became Ted's most sacred nature sites. The first was a hidden cavern near Brammo Bay that had been caused by a meteor fall, and which Tom called *Coobee Cotanyou*, or Falling Star Hole. A friend of Ted's, the writer and naturalist Charles Barrett, described it lyrically as "a cave whose mouth is overhung by ferns and jungle vines, and the lintel green with moss, a filter for water that falls upon rocks tufted with orchids the colour of dull gold."[25]

Barrett was even more impressed by a similar grotto on nearby Bedarra Island. It was hidden ten yards above the watermark on the weather side of the island, and named by Ted the Cave of Swiftlets. Within it hung more than fifty nests of gray-rumped swiftlets, each glued to the rocks with bird saliva and containing a "pearly white egg." These rare little birds, first seen by Jock MacGillivray of the *Rattlesnake*, had remained virtually unknown to whites for sixty-one years, until Tom brought their cavern to light.[26]

And only Tom knew the whereabouts of two legendary rock-art galleries on Dunk, which had long been lost within the island's mountainous and overgrown rain forest, even to other Aborigines. Ted had little understanding of art, and his descriptions of the red and brown ochre rock paintings were laced with condescension, yet he was moved to write: "Here is the sheer beginning, the spontaneous germ of art."[27]

In time Ted grew, almost despite himself, to venerate Tom and Mickie. Here were men who could swim huge distances without fear of sharks, who could sail the crankiest of craft in any sea, who could spear fish and turtle with preternatural speed and accuracy. They could improvise traps and nets capable of snaring fish of all sizes and speeds. They could spot tiny objects at a distance and with a clarity that exceeded the range of binoculars. Like the Kaurareg, they could catch a two-hundred-pound bull turtle using a remora suckerfish that clamped onto the creature's shell. They knew the art of stunning fish by crushing an array of "wild dynamite" plants, in a process so recondite it proved to Ted that "the Australian aboriginal has to his credit as a chemist the results of successful original research, and . . . he is also a herbalist from whom it is no condescension to learn." There was a note of awe in his praise:[28]

Mickie's bush craft, his knowledge of the habits of birds and insects and the ways of fish, is enviable. Signs and sounds quite indeterminate to "white fellas" are full of meaning to him . . . The scratching of a scrub fowl among decayed leaves is heard in the jungle at an extraordinary distance, and a splash or ripple far out on the edge of the reef tells him that a shark or kingfish is driving the mullet into the lagoon, where he may easily spear them. He can tell to a quarter of an hour when the fish will leave off biting . . . and knows when the giant crabs will be "walking about" in the mangroves. He is trustworthy and obliging, and ready to impart all the lore he possesses, an expert boomerang thrower, a dead shot with a nulla-nulla, and an eater of everything that comes in his way except "pigee-pigee" [nutmeg pigeon].[29]

Mourning Tom's premature death by spearing in a mainland melee in 1911, Ted allowed his admiration to spill over into open affection. This broad-chested, big-limbed, coarse-handed warrior had been as gentle and funny as he was brave, as tender as he was tough, as learned as he was skilled. "Among his mental accomplishments was a specific title for each plant and tree," Ted wrote. "His almanac was floral. By the flowering of trees and shrubs so he noted the time of the year, and he knew many stars by name and could tell when such and such a one would be visible."

Penning an impromptu epitaph for Tom, Ted summoned his highest words of praise: "he [was] an Australian by the purest lineage and birth—one whose physique was an example of the class that tropical Queensland is capable of producing, a man of brains, a student of Nature who had stored his mind with first-hand knowledge unprinted and now unprintable . . ."[30]

Above all, it was his Aboriginal friends who led Ted to develop one of the foundational beliefs of his distinctive beachcomber philosophy: that individuals must develop "a sense of fellowship with animated and inanimate things" within their country. Such knowledge must draw on the complete spiritual, material, emotional, sensual, and intellectual composition of one's being. Dunk Island was not just a habitat or environment, it was a fusion of nature and culture: a heartland, a Dreaming. When Ted wrote of his duty "to exhaustively comprehend" his island, he was referring to the Aboriginal way of comprehending—a way of learning that any true naturalist should follow:

If you would read the months off-hand by the flowering of trees and shrubs and the coming and going of birds; if the inhalation of scents is to convey photographic details of scenes whence they originate; if you would explore miles of sunless jungle by ways unstable as water; if you would have the sites of camps of past generations of blacks reveal the arts and occupations of the race, its dietary scale and the pastimes of its children; if you desire to have exact first-hand knowledge, to revel in the rich delights of new experiences . . . [31]

If you really wanted to learn all these things, Ted concluded, then you must put yourself under the tutelage of an Aboriginal like Tom or Mickie.

Yet surprisingly, the catalyst that triggered Ted's transformation into a fully fledged author came from outside the island. On October 10, 1904, the weekly steamer disgorged a tubby, scruffy, red-faced Englishman of fifty-two who'd invited himself by letter for a two-week visit. Notwithstanding his eccentric appearance, Walter Strickland proved to be one of the most compelling individuals Ted would ever meet.

He was thrilled to learn the man's aristocratic lineage. Though Strickland dressed like a hobo, he was the son of a Yorkshire landed baronet and due to inherit his father's title, castle, and wealth. On top of this he'd accumulated impeccable literary and scientific credentials. Educated at Trinity College, Cambridge, he was fluent in several languages, had a string of publications to his name, and was well versed in natural history. Strickland's fascination with the corals, marine creatures, and bird life of the island equaled Ted's. Impervious to hardship, the Englishman spent each day scrabbling over reefs, peering into rock pools, and quizzing Ted on his knowledge of bird nesting and migration patterns. He also pressed Ted to make a systematic census of all the bird species on Dunk Island, an undertaking which eventually produced a tally of 128 species, not counting a dozen or so that Ted was unable to identify.[32]

A passionate conservationist, Strickland urged Ted to lobby the Queensland government to have Dunk and other islands in the Family group designated as bird sanctuaries. Ted admired the man's militant stance on the need to prevent the slaughter of Torres Strait pigeons, a species vulnerable to extinction. Thanks to Strickland's goading, Ted applied for official recognition as ranger of the Dunk and Family islands sanctuaries, a status that was granted in June 1905.[33]

Ted was also fascinated by Strickland's maverick values. Here was a man who'd rejected his aristocratic father's ambitions, repudiated the luxuries of a patrician heritage, traveled rough in countries like Indonesia, India, and China, and steeped himself in Oriental literature and Sufi philosophy. He also relished attacking the sacred cows of the British monarchy and empire, and of Christianity. Ted, as a conservative and imperialist, didn't share such outlandish views—and Bertha found them repellent—but he was pleased to be treated as a fellow rebel against social conformity.

Even more flattering was Strickland's genuine admiration of Ted's nature journalism. A respected author himself, the Englishman urged Ted to gather together his occasional Dunk Island pieces into a book, and he suggested a perfect title, *The Confessions of a Beachcomber*. The lure of becoming an international author rather than a mere local journalist captured Ted's imagination, and held it for the rest of his life.

Soon after departing, Strickland kept his promise to assist the book's publication and in 1906/7 Ted's manuscript was accepted by the London publishing firm T. Fisher Unwin, specialists in naturalist and travel works. Even so, the publisher's demand for a subsidy of 150 pounds would have been well beyond Ted's means had Strickland not loaned him the money. After months of frenetic writing and revision, the book eventually appeared in London on September 17, 1908, and in New York the following year.

Almost immediately, Ted began to receive a stream of heady reviews that compared him to famous literary figures like Robert Louis Stevenson and Thoreau. Ted Banfield, Dunk Island beachcomber, had joined their number. But though feted in its day, Ted's writing wasn't always good. The down-to-earth Bertha advised him to be less flowery, sensing that his mimicry of the styles of Shakespeare, Dickens, and Lamb didn't always suit the subject matter or the times. Still, when the different strands of his muse did come together he produced nature writing of genuine brilliance. At his best when describing the beauties and wonders of Dunk's fringing reefs and marine life, he combined the imaginative power of a romantic poet with the forensic insight of a scientist and the holistic understanding of an Aborigine.

Addressing his readers as intimate friends, he invited them to join him on his rowboat to see what the fringing reefs of Brammo Bay had looked like before a cyclone in 1903: "To see the coral garden to advantage you must pass over it—not through it. Drifting idly in a boat on a calm clear day, when the tips of the tallest shrubs are submerged but a foot or so, and all the delicate filaments, which are invisible or lie flat and flaccid when the tide is out, are waving, twisting and twining, then the spectacle is at its best."[34]

Ted drew attention to the tiny, seemingly humdrum marine creatures that had built this complex spectacle: "Apart from the bulk and fantastic shapes of coral structures, there is the beauty of the living polyps. That which when dry may have the superficial appearance of stone plentifully pitted—a heavy dull mass—blossoms with wondrous gaiety as the revivifying water covers it . . . Here is a buff-coloured block roughly in the shape of a mushroom with a flat top, irregular edges, and a bulbous stalk. Rich brown alga hangs from its edges in frills and flounces. Little cones stud its surface, each of which is the home of a living, starlike flower, a flower which has the power of displaying and withdrawing itself, and of waving its fringed rays."[35]

At this point it may be useful to note that coral polyps belong to the phylum Cnidaria, which includes jellyfish. Polyps as a whole are carnivorous, multicellular animals with a two-layered body plan and a mouth surrounded by fine tentacles, while coral polyps are a small, soft-bodied form with a cylindrical trunk. The term "coral polyp" has been used by scientists since the mid-eighteenth century, a result of the work of Jean-André Peyssonnel and John Ellis.

Keen to arouse a sense of aesthetic wonder in his readers, Ted nevertheless disdained any associated sentimentality, presenting himself as a naturalist committed to Darwin's great law of the survival of species by natural selection.

A coral reef is gorged with a population of varied elements viciously disposed towards each other. It is one of Nature's most cruel battlefields, for it is the brood of the sea that "plots mutual slaughter, hungering to live." Molluscs are murderers and the most shameless of cannibals. No creature at all conspicuous is safe, unless it is agile and alert, or of horrific aspect, or endowed with giant's strength, or is encased in armour . . . The whole field is strewn with the relics of perpetual conflict, resolving and being resolved into original elements. We talk of the strenuous life of men in cities. Go to a coral reef and see what the struggle for existence really means. The very bulwarks of limestone are honeycombed by tunnelling shells.[36]

Yet in mitigation of this "perpetual war of species," he cited the numerous instances of "commensal" behavior on the Reef, such as when a pinna mollusc gives lodgings to a mantis shrimp and a miniature eel in exchange for food and cleaning services.

Even symbiotic allies like these could not, however, protect the mollusc from the parrot fish, simultaneously one of Brammo's most exquisite creatures and most ruthless predators. As a naturalist, Ted wrote that this fish belonged to the scaroid family, possessed a beaklike mouth, a row of pharyngeal interior teeth for grinding hard shell, and a gizzard "composed of an intensely tough material, lined with membrane resembling shark's skin." As a romantic, he described it as the "jewel of the sea," having iridescent scales "of slightly elongated hexagons, generally blue outlined with pink, sometimes golden-yellow combined with green; and the colours flash and change with indescribable radiance." And as a disciple of the "natives of the island," he recorded that the fish's flesh was edible, though not particularly flavorsome, and that it was known by the euphonious name of "Oo-ril-ee."[37]

The two books that followed the bestselling *Confessions*—*My Tropic Isle* in 1911, and *Tropic Days* in 1918—completed a beachcomber trilogy. *My Tropic Isle* matched its celebrated predecessor, both in quality and international success. Ted met his readers' wishes for greater personal details of his joyous beachcomber life, and his publisher's requests for more of his vivid descriptions of marine life. *Tropic Days*, however, faltered: he tried to combat his exhaustion of subject matter by including fictional and quasi-fictional pieces that were hackneyed in comparison with his real-life nature writings.

There was also a deeper, mental problem. From the outbreak of World War I in 1914, Ted had begun to lose confidence in his literary creation of the insouciant beachcomber. It was not that the war initially brought many changes to Ted's life; in fact, that was precisely the problem. When so many others were enduring sacrifice and suffering, Ted felt guilty about living in an escapist paradise. An ardent patriot, easily enraged by newspaper reports of German atrocities, he was also genuinely depressed by the black-rimmed lists of Australian casualties. Letters to family confidants like his sister Eliza hinted at a fear that he and Bertha would somehow be made to pay for their carefree island life.[38]

Nemesis, as he was later to think of it, arrived with two monster storms. The first reached Mackay to the south on January 21, 1918, the second hit Dunk Island and the surrounding region on Sunday, March 10. As with all imagined apocalypses, Ted later recalled strange portents: a change in the quality of heat, from moist steam bath to scorching furnace; flocks of disordered frigate birds wheeling aimlessly in the sky; and "Wan Tam snakes congregating around the poultry yard." Then, between 9:00 and 10:00 p.m., "with a conglomeration of terrifying sounds varying from falsetto shrieks to thunderous roars, the center of the cyclone seemed to bore down on the very vitals of the island." Ted, Bertha, and an Aboriginal family from the settlement at Mission Beach dashed from room to room as the bungalow's tin roof peeled off, wheeling away into the dark, until only Ted's study remained intact. The little group of refugees crouched on the floor while trees splintered around them and the contents of the house swirled, as if "crazy with palsy."[39]

The cyclone blew for a manic half hour, lapsed briefly, then resumed with a last burst of violence before disappearing. The next day Ted learned that the nearby town of Innisfail had been wrecked and fifteen lives lost. Even closer to home, his friend John Kenney, Superintendent of the Mission Beach Aboriginal Settlement, had been killed while trying to rescue his daughter, who was actually already dead, a spear of broken timber driven through her heart. At least three Aborigines had been drowned and some two hundred others were missing.

Surveying Dunk Island itself, Ted saw a scene of carnage, "a leafless wilderness" that aroused images of the Somme: "The shrub-embroidered strand is now forlorn, its vegetation, uprooted and down-beaten, naked roots exposed to critical view. Not a shrub has escaped, and broken and shattered limbs of tough trees appeal for sympathy." He couldn't help using words like "fouled" and "soiled" to describe his lost paradise. Dunk had been ravaged by a "great, bullying" nature that he'd previously described with metaphors of purity and innocence.[40]

Every tree was uprooted or "maimed and disfigured." The streams were "foul-tasting" and "blocked with decaying vegetation"; the islet in the bay, once dotted with scarlet umbrella trees and golden-brown orchids, was now a bare rock. Ted and Bertha's grand avenue of coconut palms had been tossed fifty yards away, and a consequent tidal wave had strewn the garden area with mounds of sand, coral, and shells that extended one hundred yards beyond the strand. Their new motorboat had been torn from its trolley and pulverized. Worst of all, thousands of birds had been killed: as if mown down by machine guns, their corpses lay among a sea of rotting fruit and debris. Ted wrote despairingly to his Ararat relatives that his and Bertha's life at Dunk was over. Both being in their sixties, the task of starting again seemed impossible.

As usual, it was Bertha who rallied Ted's spirits. Calling him over to some smashed fruit trees, she pointed to fresh buds and leaf shoots. "So let's get the clocks going again," she said briskly. Within a week they'd decided to stay. Mainland friends rallied around with gifts, labor, and promises of money. Bertha reminded Ted, too, about his pontifications in *Confessions* about nature's resilience. Encouraged, he began to see signs of hope and renewal. The sun, able to penetrate into new-made forest clearings, was germinating bird-carried seeds to create a fresh Elysium: "instead of permanently destroying vegetation, the big wind will have to its credit denser and more beautiful growths; instead of grassy glades, an almost impenetrable entanglement; palms will sprawl over lofty trees; huge vines, with stems as thick as a man's thigh and bearing pods a yard long, will spread a network over all; and instead of the forest's comparatively dry surface will be maintained a moist, sweet-smelling soil, and steamy conditions and half-lights."[41]

One day Bertha called Ted's attention to the most redemptive sign of all—the return of a pair of sunbirds, without which "the Isle would have lost no little of its glitter."[42]

As in the past, though, Ted experienced a type of delayed shock. Bertha noticed that he was oscillating between bouts of feverish elation and manic overwork. Recognizing the signs of an impending nervous collapse, she ordered him to take a week's rest in the Townsville hospital. This, combined with the news that a friend had offered them a fixed income in exchange for a half-share of their land, brought Ted back to relative normality.

In the longer term, though, he never recovered from the combined effects of the war and the cyclone. They seemed to instill in him a sense of disquiet as profound as that of many veterans returning to a world they no longer understood. Among the sources of Ted's postwar malaise, one irritant proved especially galling, because he himself was to blame. The fame of his beachcomber books and articles brought a tide of inquiries from romantic-minded Australians, Europeans, Britons, and Americans who wanted "to go a-Dunking." Most he rejected, but some turned up anyway. And a few had war records or personal disabilities that touched his susceptible heart. He and Bertha accommodated several struggling families on the island during the early 1920s, though the couple invariably found the experience testing. Other visitors, like the Governor of Queensland, were too grand to turn down, or, like the cartographers and scientists from the survey ship *Fantome*, were engaged in work too important to refuse.

Most troubling were all the new entrepreneurs and tourists who seemed to be invading the Reef region: vacationers in cars, steamers, and yachts; pearlers, trochus-shell collectors, and trepang fishermen in praus, luggers, and ketches; and gun-happy Italian sugar workers or soldier-settlers who wanted to open up new stretches of the coastland. Ted didn't actually oppose the development of northern Queensland; as the author of two tourist guides extolling the scenic, climatic, and economic attractions of the Reef, he could hardly complain. But he was infuriated when these new people came as plunderers and killers. Some collected "bird skins," dugong oil, and turtle meat for money; many killed pigeons and starlings for sport; and all collected corals and shells indiscriminately.[43]

Ted compared Dunk Island birdlife in 1921 to his census of 1905. The results were shocking: every species of bird of prey had vanished, as had fruit pigeons, oystercatchers, plovers, and egrets. All other bird species, except the brown-winged tern, showed a substantial decrease in numbers. True, the cyclone was partly to blame, but it was nutmeg pigeons and metallic starlings—the species most attractive to sportsmen—that had suffered the most.[44]

Ted was convinced that the Queensland government bodies responsible for wildlife protection were turning a blind eye to this destruction. They issued shooting licenses promiscuously, and refused to restrict shooting seasons for birds. Ted wrote searing newspaper denunciations of "selfish collectors" and the "traffic in wild birds," but the final straw was receiving a government poster advertising a new Animals and Bird

Act of 1921 that made no mention of Dunk Island or the rest of the Family group as bird sanctuaries. One cavalier pen stroke by a Brisbane bureaucrat had excised twenty years of work toward Ted's "grand objective." He wrote a furious letter resigning his cherished position of ranger, a gesture that the department didn't bother to acknowledge.

Some consolation came from his association with a few like-minded men and women who were equally committed to the preservation of Dunk Island's bird and marine life. Alec Chisholm, an influential naturalist, conservationist, and journalist, became a close friend to both Ted and Bertha, fighting in the popular press for many of the beachcomber's causes. (Bertha, for her part, seems not to have considered herself a beachcomber, being more conservative than Ted; she went along with him in a spirit of spousal tolerance.) In September 1922 Chisholm reported triumphantly in the *Daily Mail* that, thanks to Ted's passionate advocacy, the government had finally agreed to the complete protection of nutmeg pigeons.[45]

The glow of this vindication had no time to fade before Ted suddenly succumbed to peritonitis, dying peacefully in the night of June 2, 1923, at the age of seventy-one. Bertha, isolated by storms, had to stay with his body for three nights before managing to attract the notice of a passing steamer. The captain and sailors came ashore, built a rough coffin of ship's timber, and laid the beachcomber to rest in the island he so loved. Later Bertha built a cairn over his grave; she eventually joined her "dear laddie" there in August 1933.

Ted Banfield left many legacies, including the collection of late writings published by Alec Chisholm in 1925 under the title *Last Leaves from Dunk Island*, to favorable reviews from all around the world. Yet perhaps his greatest legacy was the propagation of a new, more respectable version of the island paradise myth, devoid of fantasies about Polynesian maidens. The idea of the island paradise had until now been associated primarily with the South Seas, and Ted's beachcomber paradise helped put an end to lingering assumptions that the Barrier Reef was full of violent, primitive headhunters.

Ted was responsible for the idea that the Reef in fact contains multiple island paradises. His books, which were far more widely read than William Saville-Kent's expensive work, generated a spate of would-be Crusoes, and in the long run added to interest in the possibility of Reef tourist expeditions and resorts. It was a legacy that Maurice Yonge and his young biologists would consolidate and extend.

Given Ted's desire for Dunk to become a wildlife sanctuary, its transformation into a resort is ironic. Still, we can imagine him smiling grimly to learn of the latest act of Nemesis, in 2011, when Cyclone Yasi all but blew the Dunk Island resort away.

PART THREE

Wonder

9

OBSESSION

The Quest to Prove the Origins of the Reef

PEOPLE WHO KNEW American zoologist Alex Agassiz described him as a kindly, gentle, and eminently rational man, yet he spent much of the latter part of his life engaged in what appears to have been an obsessive, expensive, and quixotic quest to visit and analyze all the coral reefs of the world. His epic series of Pacific and Indian ocean voyages began on April 16, 1896, with an expedition to the greatest reef of all.

Agassiz was already sixty-one when he embarked on the small chartered steamer the *Croydon* from Brisbane. One of his assistants on that trip, Alfred Mayor, recorded that during the next twenty-five years, Agassiz would "wander further and see more coral reefs than has any man of science of the present or past."[1]

Alex Agassiz's aim was ambitious: he wanted to discredit Charles Darwin's famous theory of the origin of coral reefs, and where better to start than Australia. His *Croydon* expedition marks yet another change in the perception of the Great Barrier Reef, which, having experienced varied shades of fame and infamy, was catapulted by Agassiz's quest to global scientific prominence. The Reef became a centerpiece in a fierce debate among the world's leading geologists, marine scientists, and oceanographers—a debate that was as important internationally as it was in Australia.

Why, though, should an aging and reclusive expert on starfish like Agassiz choose this moment to mount such an obsessive mission to the other side of the globe? It was equally odd for Charles Darwin's theory of coral reefs to be generating such heated debate so long after its first drafting. After all, this had been Darwin's juvenile scientific achievement, scribbled sixty years earlier when voyaging around the world as a young gentleman's companion on the HMS *Beagle*. He confessed that the theory had been an act of pure deduction, having come to him when he was scrambling up the Andes Mountains in Chile, before he'd ever seen a coral reef. Coral fossils within the Andes rock strata suggested to Darwin that this vast mountain chain had been elevated gradually from the ocean floor by volcanic action over aeons of time.

But what goes up can also go down: perhaps, thought Darwin, a corresponding subsidence had taken place on the bed of the Pacific and Indian oceans? This would explain a mystery long puzzling to navigators and naturalists: how these tiny coral "insects," known to live only in shallow, light-filled waters, had built vast ramparts that rose up from the dark depths of the oceans.

If, speculated Darwin, coral reefs grew in a ring pattern around the shallow fringes of volcanic pinnacles in the sea, and if these rocky islets were to subside at a pace that matched the reefs' upward growth, they would create thick walls of dead coral with a living crust on top. Over many centuries this process would leave behind circular atolls surrounding shallow lagoons, or barrier reefs separated from the mainland by lagoon-like channels.

Charles Darwin's few hasty observations in the Pacific and Indian oceans seemed to confirm the theory, which was then strengthened over many years by the closer investigations of James Dwight Dana, a former Pacific voyager who became an eminent Yale geologist. Since then few scientists had challenged this simple and elegant hypothesis; it was one that Darwin, in his autobiography of 1876, regarded as "well established."[2]

Alexander Agassiz was born in Neuchâtel, Switzerland, in 1835, the year that Darwin first drafted his reef-subsidence theory. His father, Louis Agassiz, migrated from Switzerland to America in 1846, where he quickly became a Harvard professor of zoology who dazzled everyone he met, scientists, literati, and glitterati alike. His European charm, dramatic lecturing style, and romantic personality helped make him the first American scientist to stride the global stage since Benjamin Franklin.

But Louis still believed that natural species were God's thoughts ordered in a beautiful "Plan of Creation," and following the publication of *On the Origin of Species* in 1859 it was inevitable that he would clash with Darwin and his supporters. A decade on, Louis Agassiz's increasingly strident and ineffectual attempts to snuff out Darwin's theory of evolution had cost him much scientific respect.

Most of his son Alex's troubled early childhood was spent either at boarding school or living with his artistic mother Cecile in Freiburg, Germany. She had taken her children there when she could no longer stand living with her narcissistic, work-obsessed husband. After a few years she died of tuberculosis. This forced Alex, at the age of thirteen, to join Louis and his new stepmother, Liz Cary, in Boston. Fortunately, Alex doted on them both. Alfred Mayor later claimed that the son's "reverence for his father was almost a religion with him."[3]

On graduating from Harvard in engineering, zoology, and natural history, Alex spent most of the 1860s trying to control his father's feckless spending of the budget of Harvard's new Museum of Comparative Zoology (MCZ). Deeply sensitive under a reserved exterior, Alex also had to undergo the anguish of witnessing his father's successive scientific humiliations at the hands of Darwin and his prominent Harvard supporters, professors Asa Gray and James Dana. On top of this, Louis's high-handed manner alienated a bright group of students, who seceded from the museum with much scandal after accusing him of plagiarizing their work.[4]

Alex Agassiz was no cipher, however. Louis's opposite in personality, this shy, meticulous boy was determined to make his own way in life. Having fallen deeply in love with one of his students, the independent-minded Anna Russell, he married her in 1860. The devoted couple had three children in quick succession and forged a close-knit circle of half a dozen wealthy Boston relatives and friends. Six years later Alex surprised everyone by deciding to rescue a struggling Michigan copper mine, which he miraculously turned into one of the largest and most prosperous in America, making him a millionaire.

Continuing at the same time to exercise his passion for marine biology, Alex employed methods that were as thorough as his father's were cavalier. During the 1860s and '70s he produced an impressive list of publications on the taxonomy and embryology of echinoderms (the phylum which includes starfish, sea urchins, sea cucumbers, and others), all of which displayed a rigorous empirical approach to evidence. Privately he also admitted to a "general" acceptance of his father's bête noire, Darwinian evolution. On a trip to England in 1869 Alex even took the trouble to meet and impress both Darwin and his disciple, Thomas Huxley.[5]

But in mid-December 1873 Alex Agassiz's world collapsed. "The thunderbolts of God fall heavily upon us," wrote his best friend and brother-in-law, Theo Lyman. The first shock was when Louis Agassiz died suddenly on December 14, after suffering a massive stroke. A few nights later Alex's beloved wife, Anna, exhausted from tending to her father-in-law, was diagnosed with pneumonia. Aged only thirty-three, she died at midnight on the twenty-second and was buried on Christmas Eve. "Alex stood at the brink [of the grave]," Lyman recorded, ". . . with the tears rolling down his face, till I whispered to him to go."[6]

All life washed out of Alex; he couldn't even comfort his distraught sons. "I am utterly unable to get reconciled to an existence which is well-nigh intolerable," he confessed to the German biologist Ernst Haeckel. Six months later, while trying to cover his father's teaching obligations, he broke down "and cried without control; and seemed like a man who'd lost much blood."[7]

As months and then years passed, the grip of this depression showed no sign of lessening. Frenetic work and a series of overseas trips offered some distraction, but Alex's close friends, with whom he was often melancholy and withdrawn, noticed permanent changes in his personality. He became gruff, surly, and prone to explosive bouts of anger with his students and employees. His young assistant Alfred Mayor commented that Agassiz "raised a wall between himself and the unsympathetic world . . . he held himself far and aloof."[8]

In late 1876, still as fragile as ever, Alex accepted an invitation to visit Britain to help the celebrated Scottish marine scientist John Murray sort through some of the thousands of specimens of animals, plants, and seafloor deposits collected on the HMS *Challenger*'s three-year oceanographic expedition around the world. The meeting started awkwardly:

Alex silenced Murray's opening commiserations about Anna with the blurted cry, "I cannot bear it." Yet working alongside Murray he grew animated for the first time in years, as he learned of the expedition's startling results. These included the discovery of teeming life in the supposedly barren oceanic depths, and the mapping of vast gullies, canyons, and mountain peaks on the seabed.[9]

But Alex was intrigued most by Murray's revelation that billions of calcium skeletons from minute, single-celled plankton rained ceaselessly down onto the ocean floor. This suggested, said Murray, that Darwin's subsidence theory was not needed to explain how atolls and barrier reefs had come into being. It was possible that, given a perpetual avalanche of dead plankton tumbling through the ocean depths faster than it could be dissolved by the carbonic acid in seawater, this massive detritus had settled on the numerous rocky mounds already pushed up from the ocean floor and then amalgamated into sedimentary platforms. Eventually these limestone platforms would have reached a height close enough to the surface light for corals to begin growing.

Once these corals reached the ocean's surface, the violence of the breakers would create a further base, or talus, of eroded rubble and broken corals. Toward the windward edges of these, a patina of living corals would flourish by feeding on the wave-carried plankton, but those corals sheltered to leeward would starve and die, their calcium skeletons gradually dissolved by seawater. Thus crescent-shaped atoll lagoons or canal-like barrier lagoons would be formed, depending on the original shape of the base.[10]

Alex Agassiz returned to America barely able to contain his excitement: "It is the first time since the death of my father and my wife that I have felt in the least as if there were anything to live for," he wrote to Wyville Thomson, the leader of the *Challenger* expedition. Up to this time he'd thought of himself as a marine zoologist, leaving issues like the origins of coral reefs to the geologists. But the scales had fallen from his eyes, and although he'd earlier agreed with Darwin's coral reef theory, he now denied it. "I never really accepted the theories of Darwin," he told John Murray. "It was all too mighty simple."[11]

What especially troubled Alex about Darwin's theory was that subsidence hid the evidence of its operations, and seemed almost impossible to prove. Murray's alternative explanation was both multifaceted and testable. Here was a wounded son suddenly offered the chance to re-venge his late father's humiliations, and Alex grabbed the opportunity with alacrity.[12]

A further goad was awaiting him on his arrival in America, in the form of a new publication from Darwin's most fervent German disciple, Ernst Haeckel. Alex had always thought of Haeckel as a close friend, one of the few people to whom he could confide his agonies of personal grief. What he now read shocked him to the core. Haeckel had written a jeering, sarcastic pamphlet called *Goals and Paths*, which libeled Louis Agassiz's character and legacy under the guise of discussing recent biological trends. It accused Louis of having cringed to the creationists, of having stolen his only decent scientific idea from his colleagues, and of having being "the most ingenious and energetic racketeer in the entire domain of natural history." Alex sent off a furious letter in reply, calling Haeckel "an unmitigated blackguard" and breaking off all future relations. He told his uncle that he'd like to take a horsewhip to the man.[13]

Alex Agassiz was too fair-minded to blame Darwin directly for such a rogue disciple. Still, when undertaking a series of navy-sponsored oceanographic cruises in Florida and the Caribbean over the next few years, he began looking for evidence to support Murray's new theory. In 1880 Murray presented his case against Darwin in *Nature*, claiming to have seen archipelagoes in Tahiti, the Maldives, and Fiji with no signs of subsidence, but strong evidence of elevated sedimentary platforms.

Emboldened, Alex wrote to Darwin predicting that future reef expeditions would confirm Murray's results. Darwin didn't miss the note of challenge, and though tired and ailing he countered with one of his own: "If I am wrong," he wrote wearily, "the sooner I am knocked on the head and annihilated the better . . . I wish that some doubly rich millionaire would take it into his head to have borings made in some of the Pacific and Indian atolls, and bring home cores for slicing from a depth of 500 or 600 feet." If Darwin were indeed wrong, such cores would show a superficial crust of coral, underlain by extensive older submarine rock. If he were right, the cores would show a considerable depth of coralline limestone.[14]

Darwin died less than a year later, which is probably why Alex didn't at once take up the challenge, aside from making a short, inconclusive tour of reefs in Hawaii in 1884. The following year, however, he wrote to James Dana at Yale, endorsing Murray's arguments and outlining a personal "dream" to hire a vessel to investigate Pacific islands and reefs "with modern methods," so as to solve this compelling geological problem. Dana's reply was less gentle than Darwin's had been: he published a long paper pouring such scorn on Murray and his supporters that Alex immediately broke off all relations with the man.

One last inducement pushed Alex Agassiz into a crusade to prove Darwin wrong: a public attack against the Darwinists mounted by the Duke of Argyll and a trio of eloquent English bishops. They accused "the Darwin faction" of mounting "a conspiracy of silence" and "reign of terror" to stifle recognition of Murray's theory of coral reefs, and to mask the fact that Darwin's own "errors [were] as profound as the abysses of the Pacific." Huxley, famed for writing with vitriol rather than ink, seared the group with his customary brilliance, but the Argyllites were not easily quelled.[15]

Their tactic of diverting attacks away from *On the Origin of Species* and onto Darwin's theory of coral reefs was a shrewd one. There were obviously close links between the two theories: Darwin's *The Structure and Distribution of Coral Reefs* (1842) had anticipated its famous successor in both form and content, and it had launched Darwin's arc of scientific fame. Disproving the coral reef theory, thereby discrediting Darwin and his disciples, would weaken the whole case for evolution by natural selection. And the fact that the reef theory was so difficult to prove made it vulnerable to demolition. In effect, Argyll and his supporters were elevating the coral reef problem into "one of the most prominent and explicitly controversial in science."[16]

The controversy over Darwin's reef theory revived the torrid evolutionary debates of the 1860s. This time, however, Alex Agassiz would not remain silent. His banner would be scientific empiricism, his field the coral reefs of the world. In 1896, quiet Alex Agassiz put to sea to revenge his father, conquer Darwin, vindicate Murray, and unlock the secrets of the coral reefs of the Indo-Pacific.

Alex Agassiz would later admit privately that his Great Barrier Reef expedition of 1896 was something of a flop, because he'd followed the advice of William Saville-Kent to visit northern Australia in April–May, which proved an unsuitable time of year. Others had recommended waiting until September–October, when the still conditions would have

Alex Agassiz (far right), Alfred Mayor (second from right), and
William Woodworth (third from right) on board the *Croydon*,
1896 (Archives of the Museum of Comparative Zoology, Ernst Mayr Library,
Harvard University)

been ideal for reef viewing. As it was, Alex, his son Maximilian, and two
young Harvard museum zoological assistants, William Woodworth and
Alfred Mayor, faced almost two months of buffeting southeasterly trade
winds that whipped up a choppy sea and forced their ship to remain in
harbor for all but three days of their two-month visit.

The *Croydon* was also, Mayor wrote in his journal, "a plebean [sic]
little tramp steamer only 180 feet long and she floats so low in the water
that the waves wash over her deck in a most disrespectful manner." The
American scientists were confined to inspecting the southern inner portion of the Barrier Reef region, between Breaksea Spit and Lizard Island.
Mayor reported that they managed only two fleeting glimpses of the
outer Barrier—"the grandest coral structure in the world"—and even
these were from "a respectful distance."[17]

This was all the more frustrating because of the care with which
Agassiz had planned the expedition. The *Croydon* carried "a complete

photographic apparatus and an extensive outfit for pelagic fishing," as well as deep-sea nets and a sophisticated sounding apparatus built specially by a U.S. naval engineer. Before setting off, Agassiz had studied every available Barrier Reef chart, explorer's account, and scientific paper. These ranged from the pioneering works of Flinders and Jukes, to Saville-Kent's great work of 1893, *The Great Barrier Reef*, and several recent geological analyses by Australian scientists Charles Hedley, Thomas Griffith Taylor, Ernest Andrews, and Edgeworth David.[18]

Reading Jukes and Saville-Kent was particularly important because they'd produced the most comprehensive Reef surveys, and, still more, because both had endorsed Darwinian subsidence. Alex conceded that Saville-Kent's "superb" photographic plates were unique in giving "an idea of the appearance of a coral reef," but he thought the man's conclusions were weakened by his "writing in the popular manner." Jukes's analysis was so good, he admitted, that little could be added to it. Yet Alex thought that Jukes had ultimately come to "erroneous conclusions" because he'd allowed "his admiration for the simplicity of the explanation of the theory of coral reefs by Darwin to blind him."[19]

Specifically, according to Alex, Jukes had failed to see the significance of "the mass of islands that crop out nearly all along [the Reef]." He had assumed that these were originally mountains on a fragment of the mainland that had then subsided under the sea, but Alex believed the islands had been elevated from the seabed and eroded by waves, wind, and rain until they were leveled into rocky flats. Corals had subsequently grown on top of them in a thin veneer, which Alex guessed would be no more than ten to twelve fathoms thick.[20]

Alex also had to deal with a strong consensus among Australian geologists that extensive ocean-floor subsidence had taken place during the Cretaceous period, some sixty-five million years earlier, breaking up a larger Pacific continent and leaving Australia behind. Alex didn't dispute this idea, but he did think it ludicrous that there could be any connection between the present-day Barrier Reef, which he believed a relatively modern production, and this ancient subsidence. If Australia's Barrier Reef had begun growing in that remote period, the corals would "have a thickness which should correspond to a depression of at least 2,000 feet." Such an unimaginable thickness of coral was, he believed, too absurd to need refuting.[21]

Alex claimed publicly that even his abbreviated investigation of the

Great Barrier Reef had proved the essential correctness of Murray's theory. "I began to have my eyes opened, and to get an explanation of the formation of the coral flat reefs," he wrote after surveying two reef patches at Lark Passage near Cooktown on May 5. Confirmation, he thought, came a week later when he was exploring reefs at Hope Island, not far from Cairns. "Here," Mayor recorded, "greatly to Mr. Agassiz's joy he found the reefs so thin that he actually obtained specimens of the granite rock under which the coral grows. This settles the question that the reef is formed in the Murray manner and not in that suggested by Darwin . . ."[22]

Persuading himself that he had, despite the Reef trip's vagaries, essentially confirmed his hypothesis, Alex next decided to take on Fiji, using much the same team as before. This time he arrived in mid-October, when conditions suited reef viewing. Just before reaching Suva, however, he heard disturbing news. Britain's Royal Society had commissioned a party of Australian scientists and technicians, led by the University of Sydney's Edgeworth David, to undertake a deep drilling of the coral reef at Funafuti, north of Fiji. They'd driven down through six hundred feet of limestone before the drill gave out. Although by no means conclusive, this appeared to favor Darwin's idea that the coral had thickened as the ocean bed sank.

Pushing this uncomfortable news aside, Alex's team proceeded again to fight against the Darwinists. Everywhere in Fiji that James Dana had seen subsidence, Alex saw elevation "of at least eight hundred feet." Most of the corals, he said, were growing on beds of elevated volcanic lava, although a few reef platforms were composed of what he called "old marine limestone." This, he stressed, was not coralline limestone as predicted by Darwin, but a more ancient, finer-grained "marine . . . sedimentary rock composed of the remains of zillions of tiny sea animals," and elevated from the bottom of the seabed.[23]

To his delight, Alex later received some encouraging news from Edgeworth David, with whom he was corresponding. After drilling through forty feet of coralline crust, the Funafuti scientists had also noticed a different type of limestone in their cores, which Alex took as evidence of elevation. His confidence swelled: "I shall give them [the Darwinists] a dose they do not expect," he wrote to Murray in triumph, "and the theory of subsidence will, I think, be dead as a doornail and subside forever hereafter." Darwin's Fiji observations, he told another friend, had come from studying charts in his house: "a very poor way of doing, and that's the way all his coral reef work has been done." It seemed absurd to him that the subsidence theory had "got such a hold with so little holding ground."[24]

Alex's subsequent explorations of other Pacific Island groups in 1899–1900 added incremental confirmations to what was now his entrenched interpretation. Niue, Tonga, and the island groups of the Marquesas, Paumotu, Society, Cook, Ellice (Tuvalu), Gilbert, Marshall, Caroline, and Ladrones all became a roll call of victories over Darwin and Dana. On January 18, 1902, after an expedition to the Maldives in the Indian Ocean, Alex's report to Murray bristled with contempt for Darwin: "Such a lot of twaddle as has been written about the Maldives. It's all wrong what Darwin has said, and the charts ought to have shown him that he was talking nonsense." Having by now visited all the reefs of the Pacific and Indian oceans, Alex seemed in a position to deliver Darwin a killer blow.[25]

If Alex Agassiz's campaign against Darwin had something of a mythic element to it, he failed to notice that his assistant Alfred Mayor was starting to become similarly obsessed: about Alex.[26]

Though Agassiz was fifty-seven and Mayor twenty-four when they first met, they shared some deep affinities. Both had grown to adulthood in the shadow of domineering fathers, both had lost their mothers at an early age and been brought up by devoted stepmothers. Oddly, both had also trained as engineers as well as zoologists. Mayer senior (his son changed his last name to Mayor) forced Alfred to graduate in engineering and physics before eventually allowing him to study zoology at Harvard. Unusually, too, both Alex and Alfred were talented marine illustrators. Alex had inherited his artistic leanings from his German painter mother, while young Alfred's were said to have come from the Mayers' French lineage.[27]

While still a PhD student, Alfred Mayor had stunned Alex with a beautiful sketch of a jellyfish done in Agassiz's Newport marine laboratory during a Harvard summer school session in 1892. Alex was so impressed by the drawing that he invited the young man to collaborate on an illustrated work on Atlantic-American medusae, and soon after, he offered Mayor a temporary position as a curatorial assistant of marine radiates at the Harvard Museum of Comparative Zoology.

Alex liked the ambitious spirit of this boy with a square, determined face and an artistic eye, and Alfred, for his part, was awed by the patronage of the millionaire scientist, who not only looked like Count von Bismarck, but also exuded the same gruff authority. Throughout the hot summer of 1892, Alfred and a small group of students traveled each day by stagecoach to work at Alex's private laboratory, from early morning until five o'clock. The rustic, vine-covered building was set on the slope of the shore and overlooked a private cove, from which the students could see Newport Bay to the north and the ocean to the south. "The laboratory," Alfred later wrote, "was excellently equipped with reagents, glassware, and large tanks provided with running salt or fresh water. The microscope tables were set upon stone foundations to avoid vibration, and a good little steam launch lay at her moorings . . . ready to dredge in the service of science." Afternoon swims and evening boat tows in search of jellyfish completed his pleasure.[28]

Yet despite the similarities in the two scientists' backgrounds, differences of personality, age, and circumstance began to oppress the younger man. Though he had opted for a scientific career, Alfred Mayor was, a later colleague thought, "of a distinctly artistic and poetic temperament." He nursed a vein of romanticism, a love of solitude, a Thoreau-like intoxication with the natural world, and a tendency to engage in self-conscious literary introspection, traits which had been nurtured by wandering as a boy in the woods of Maplewood, New Jersey, with his butterfly net. "I threw myself heart and soul into a world of the imagination wherein I lived apart from man, and sought my playmates among the creatures of the woods and fields. I literally loved individual butterflies I had raised from early larval stages, and exulted in their imagined joy as they flew from my hand to flutter over the clover-laden fields."[29]

The latent clash of sensibility between Alfred Mayor and his employer first surfaced on the 1896 Barrier Reef expedition. While Alex chafed and fumed at the way the trade winds were hampering his exposé of Darwin's faulty geology, Alfred fell in love with the Reef's wildlife. He'd been reading Alfred Russel Wallace's accounts of butterfly hunting in the forest clearings of Malaya, and thought himself a modern counterpart of that great explorer-naturalist.

On April 25, for example, while the *Croydon* was anchored off Dunk Island, Alfred caught a whaleboat to the beach. It was exactly six months before Ted Banfield would visit the same spot, to be smitten by the clouds of gorgeous butterflies and the bird-filled glades of native forest. Mayor anticipated the beachcomber's sentiments exactly.

> Surely nothing can exceed the luxuriant beauty of this great tropical forest and nowhere upon the Earth can we find so many shades of green in the foliage as one finds here . . . Tree ferns with their dark trunks and graceful spraying crown of leaves, all emerald green . . . Dark green deeply cleft-leaved Breadfruit trees. Palm leaves that rustled as if alive . . . Acacias, ironwood and mangroves, and giant Eucalyptus trees with their sombre slaty-green foliage standing out in sharp contrast to the rich dark-greens and yellow-green of other trees. Grass waist high covered the ground and long thin rope-like creepers hung in festoons from haunches of the ancient trees or twined in snake-like folds among boughs above.[30]

A week later, when viewing some exquisite coral formations at Turtle Reef, off Cooktown, Alfred was again moved to pen beachcomber-style reflections about the Reef's blend of wonder and terror:

> As one gazes down through the deep turquoise depths and sees the lovely play of color that the sunbeams revel in among the branches of the coral forest where shadow vies with sunshine to enchant the beholder's eye, it seems another world far from this earthly realm of ours. A place far removed from the struggle of the upper world, and where sorrow is unknown and life goes on forever in listless, languid happiness and beauty. But how different the reality of it all, for the softest fringes of the tentacles of the sea anemones and polyps that seem so beautiful in richness of color and graceful delicacy of form, are deadly stings always waiting for the unwary fish that may swim carelessly within their reach. And the beautiful fish themselves . . . are many of them cannibals, and others are deadly poisonous.[31]

Cannibal struggles were not, however, confined to the Barrier Reef. Soon after returning from Australia, Alfred Mayor became engaged to the daughter of Alpheus Hyatt, his former Harvard paleontology professor. Harriet Hyatt was a feisty and talented artist with radical views on free thought, gender equality, and female suffrage. Alfred shared her ideas, but told her that Alex was an inveterate conservative: freedom of thought and social equality were entirely absent from the Harvard Museum. He even doubted that Agassiz would stomach their engagement. Alfred claimed he dared not show passion for anything other than coral reefs and medusae. As his love for Harriet bloomed, so did his paranoia. He warned Harriet never to send any letters care of Agassiz in case the latter should guess their relationship. Bitterly the couple decided they had no choice but to keep their engagement secret.

Alfred worried that his potential promotion was also at stake. Having recently been awarded his doctorate, he was hoping for a permanent position at the museum to improve his miserly pay of one thousand dollars per year. But on top of Alex supposedly being a "huffy" and avaricious "Old Moloch" who couldn't be trusted, he was favoring another assistant at the MCZ, William Woodworth. On their Barrier Reef expedition "Little Billy" Woodworth had, Alfred claimed, fawned over Alex in a "clownish fashion," in the hope of succeeding him as director of the museum. Of course, he told Harriet, "the aging autocrat" had lapped up the flattery.[32]

Harriet, furious at both the delayed marriage and her fiancé's forced absences, blamed "Alex the Terrible" for everything, and she urged Alfred to sever ties with this tyrant who treated him like "a bondsman." She warned him that the "spider-like" Agassiz would "absorb all the men who fall in his great web": he was already gobbling up all the glory for Alfred's work.

Privately Alfred thought his patron to be reasonably generous about work attributions, but he acceded to her wish by twice drafting letters of withdrawal from Alex's relentless program of reef expeditions. Both times, though, he lost his nerve at the last minute. After all, he pleaded to Harriet, any job was better than none. But then, in August 1898, Harriet passed on the crushing news that Alex had indeed appointed the loathsome Woodworth as his successor at the MCZ. "He cares nothing for me personally," Alfred responded bitterly, ". . . our relation is only a scientific alliance."[33]

Even their scientific alliance was proving shaky. Alfred Mayor didn't share Agassiz's late-blooming antipathy to Darwin, and he disliked having to avoid mentioning the subject of Darwinian evolution in conversation. Alfred's early work on butterflies had made him a strong and continuing proponent of natural selection. He was later to write that

Darwin had done for the natural sciences what Newton had done for physics and mathematics. And though he occasionally criticized aspects of Darwin's subsidence theory of coral reefs, he was privately even more skeptical about Agassiz's alternative claims. He disagreed particularly with what would prove to be Alex's most lasting contribution to reef theory—his claim that the dissolving of limestone by fresh water could create atoll lagoons.[34]

Alfred also found his patron's views on marine biology exasperatingly old-fashioned. True, he himself respected taxonomy and was proud to have contributed toward the systematics of medusae, but he disagreed strongly with Agassiz that it should be the zoologist's only function. Trained in physics, chemistry, and mathematics, Alfred shared the view of a new breed of young American scientists that these three disciplines should become part of biology's experimental repertoire. He was also powerfully drawn to the emerging science of "ecology," which dealt with the way living things interacted with one another and with the environment. Charles B. Davenport, Alfred's zoology professor at Harvard, had taught him that biology's new mission was "to consider individuals, in mass or as species, as form-units bearing the imprint of environment and adapted thereto and as constituents of faunas."[35]

Alfred was unusual, too, in linking ecology to the cause of environmental conservation. This idea, so pervasive today, was still relatively rare in the 1890s. Marine scientists at that time tended to think the new discipline's greatest use lay in enhancing the exploitation of marine resources. While still a PhD student, Alfred had been shocked at how quickly pollution forced the closure of Agassiz's Newport laboratory. For the same reason he would later campaign in the *National Geographic* magazine against "the wanton destruction of interesting animal life" and "the impending ruin of the forests" in the once pristine environments of Florida. Alex Agassiz might have shared such concerns privately, but he disliked the science of ecology, particularly since its formal originator was Ernst Haeckel, the "blackguard" he wanted to horsewhip for slandering his father.[36]

With the dawn of the new century, young Alfred Mayor exulted in at last managing to break free of Agassiz's shackles. He and Harriet were married in August 1900, after four years of delay. Soon after, Alfred took a position as curator of natural history at the Brooklyn Institute of Arts and Sciences.

But he continued to nurse a larger ambition. He'd earlier done some fieldwork at Loggerhead Key in the Dry Tortugas, part of a cluster of tropical islands off Key West, Florida. As in the Barrier Reef region, he'd been exhilarated by the key's fantasia of underwater corals: "I sail for hours . . . over the rippling waters," he enthused to Harriet, "and look far down into the recesses of the coral caverns where the cool deep shadows invite one to plunge beneath our prosaic world into the brilliant enchanted one below." Convinced that the Dry Tortugas would be a perfect spot to establish a research center for tropical marine science, he began lobbying senior figures at the Carnegie Institution of Washington to open a laboratory there.[37]

In January 1904, thanks largely to support from influential patrons, including the supposedly hostile Alex Agassiz, Alfred Mayor found himself director of the Carnegie Institution's new department of marine biology at Loggerhead Key, the first tropical laboratory in the Western Hemisphere. Along with a substantial salary, he was given a beautiful sixty-foot yacht for marine work within the Dry Tortugas, and generous funds to build the facility.[38]

At thirty-six Alfred Mayor could hardly have imagined a more perfect fulfillment of his scientific ambitions, yet he still hoped to lead the Harvard Museum, particularly after "Little Billy" Woodworth had indeed proved a flop and was demoted from the position of assistant-in-charge in 1906. When Alfred's subsequent application for the same position failed, he angrily blamed what he described as Agassiz's "insanely jealous" attitude. At the International Zoological Congress in Boston the following year, Mayor was still furious at the knockback, claiming that Alex was as "mad as a lobster," having made "a purile [sic] attack on modern zoology" in his presidential speech.[39]

Alfred, always blind to Alex's past kindnesses, was determined that his laboratory at Loggerhead Key would focus primarily on those modern elements of zoology and marine biology that Alex had supposedly disparaged. He defined the laboratory's mission as including "intensive studies . . . in the fields of physiology, ecology, heredity, evolution, animal psychology, variation, the geology and growth of coral reefs, the bacterial precipitation of limestone in tropical seas, the chemical consti-

tution of sea water . . . the coloration of reef fishes in relation to environmental influences and natural selection . . . and the ecology and
physiology of plants of the region."[40]

He also itched to expose his former boss as a scientific dinosaur. In
1907 he'd read a paper by a rising coral biologist, Thomas Wayland
Vaughan, that attacked one of Alex's reef papers. Alfred had been delighted, crowing to Harriet that the brilliant Vaughan thought Agassiz's
work "not worth the powder required to blow it to H——!" Alfred
therefore decided to entice Vaughan to bring his innovative work on
coral biology to the Loggerhead Key laboratory. After persuading the
prickly and temperamental Vaughan to work there as a visiting fellow in
1909, Mayor triumphantly secured agreement from the Carnegie Institution to fund their proposed new projects. Alex's work would be first
in the firing line.[41]

A few months later, however, on March 26, 1910, Alex Agassiz died
in his sleep while returning by sea from Britain to the United States. His
old friend Sir John Murray was saddened to hear that nobody could find
the manuscript for Alex's long-promised book rebutting Darwin's theory
of coral reefs. Alfred was secretly delighted at the news of the missing
manuscript, but resisted gloating in a long survey of Alex's life and work
for *Popular Science Monthly*. Here he adopted a magisterial tone, praising
Alex as "my master in science, and the greatest patron of zoology our
country has known," though he couldn't resist sniping at Alex's Bismarck-
ian personality and antiquated belief that physiology and laboratory experiments were "beyond the scope of zoology."[42]

Agassiz's death did not put an end to Alfred's resentment. When
Harriet was diagnosed the following year with pulmonary tuberculosis,
having suffered it throughout their delayed marriage, Alex made Agassiz
the scapegoat, implying without foundation that the man's obsessive reef
campaign against Darwin had kept Alfred from his sick wife's side.
Within a few months of Harriet entering a sanatorium, Alfred Mayor
decided to launch a crusade of his own, to prove that Agassiz and Murray
had been wrong about coral reefs.

In 1911 Alfred unveiled an ambitious new project to the trustees of the
Carnegie Institution. If their Loggerhead Key laboratory was to achieve
its true potential, he argued, it had to undertake a major global mission.
Its research needed to be tested on the Great Barrier Reef of Australia, the
greatest coral constellation in the world. Impressed, the trustees agreed.

Even the looming likelihood of war with Germany and the setback of having Vaughan pull out at the last minute didn't check Alfred's determination to mount the expedition. He persuaded Vaughan to draw up a detailed experimental program for him and he recruited able support: three other American marine biologists and the British coral scientist Frank A. Potts, from Trinity College, Cambridge.

When they eventually disembarked at Thursday Island on September 10, 1913, Mayor and his colleagues were chagrined to discover that currents had swamped the local reefs in mud and silt. Only a scattering of larger corals had been able to "raise their heads and thrive." Determined to "die rather than fail on this expedition," Alfred took advice from Australian reef biologist Charles Hedley to transfer the project to the purer waters and extensive fringing reefs of Mer Island, in the Murray group, close to the northern end of the Great Barrier Reef.

In order to sail there through the intervening maze of reefs, Alfred was forced to track down an elusive Islander captain, whose "weather-beaten face bore many a scar, and one eye seemed to have seen better days, but more perfect discipline one never saw maintained upon a schooner." Before leaving Thursday Island, the scientists were pleased to locate some of the specific corals at Vivien Reef that William Saville-Kent had photographed and measured twenty-three years earlier. In the interim, most of the species had grown at the brisk rate of around two inches per year.[43]

Mer Island was a perfect site for a temporary laboratory. The local teacher-magistrate offered his courthouse and jail for an office and storeroom, and the Islanders proved welcoming, especially when they discovered that Potts was a friend of their famous ethnographer, Alfred Haddon. Geologically, Mer had been created by the bursting of a volcano through the limestone floor of the Barrier Reef plateau. It was a manageable 9,600 feet long and 5,600 feet wide, and hadn't suffered the hurricane and silt damage that marred many reefs farther south. The island's reef flats stretched for 2,200 feet on the windward and southeast sides, and were capped by a ridge at the far edge. This created a permanent, shallow tidal basin, 1,600 feet wide, 17 inches deep, and "densely covered with one of the most luxuriant coral growths to be found in the Pacific."[44]

On the whole, the Mer reefs didn't throw much additional light on the larger question of coral reef origins. The living coralline crusts seemed geologically recent, but the character and age of the limestone platform underneath remained uncertain. Only deep core drilling could ascertain whether Darwin or Agassiz was correct on that score. However, Mayor believed he could see confirmatory signs at Mer of an exciting theory recently expounded by a young Harvard professor, Reginald Daly.

Daly argued that previous ice ages had cooled the oceans to below the survival point of reef corals, and had simultaneously dropped the sea level in tropical regions to 120 feet below the present surface, which then also exposed the reefs to wave erosion. When the glaciers later melted and the waters rose again, corals had resumed growing along the outer edges of the platforms. In short, barrier reefs and atolls had been shaped primarily by the climate-induced rise and fall of ocean levels.[45]

It was an ingenious and important idea, and one that even Darwin had missed. Ironically it was also one that Alex Agassiz would probably have welcomed. The theory that geological environments were shaped globally during ice ages had originated with his father, who'd specifically addressed glacial sculpting. Daly had simply adapted Louis Agassiz's idea to the aquatic world of reefs.

The acrimonious debate over reef origins, then, turned out to be the least important aspect of Alfred's expedition. More important, as he later boasted to Harriet, was the fact that he'd undertaken the first-ever quantitative ecological study of a coral reef by performing systematic grid counts of coral distributions, as well as extensive tests of their temperature and silt tolerances.

To his surprise, Pacific corals proved no more tolerant of increased water temperatures than did their Atlantic counterparts. In this respect, natural selection seemed not to have improved their adaptability. Among tropical corals there was also considerable variability. Close to shore, the very warm waters killed all species of Mer corals. Within the cooler midzone area of the Reef, species like *Seriatopora hystrix* outcompeted all others. Yet these in turn were unable to survive within the agitated waters and pounding breakers at the reef's edge, where several rival species managed perfectly well.[46]

Alfred Mayor also made the discovery that water temperature was a prime factor—even more important than the smothering effects of silt—in determining the survival of most coral species. At the same time, "those forms which are sensitive to high temperature are correspondingly affected by being smothered under mud, or subjected to the influence of CO_2."[47]

Like all coral scientists, Mayor didn't yet understand why sharp changes of temperature could prove so lethal to corals, though he did note the crucial related point "that corals might be nourished in some measure by their commensal plant cells." Neither did he nor any other scientist in the world yet realize the extent to which reef-growing corals were fed and powered by what Alfred thought to be an "algal infection" of some sort in their tissues. This mystery—as we shall see—was to be investigated by a Cambridge scientist, Charles Maurice Yonge, on a later expedition to the Great Barrier Reef, at which time he would praise Alfred's Mer expedition as "the starting point in the modern study of living hermatypic [or reef-building] corals."[48]

After returning to America, Alfred Mayor continued his pioneering ecological experiments on two Samoan expeditions in 1919 and 1920, where he also initiated what was possibly the first scientific study of reef corals using diving apparatus—a cumbersome, eighty-pound brass "diving hood." But observing corals more than forty feet underwater exacerbated the tuberculosis that had made a recent appearance in his lungs, at a time when Harriet's disease was in remission. On June 25, 1922, Alfred Mayor staggered from his laboratory at Loggerhead Key in a tubercular delirium. He was found by colleagues some hours later, lying dead on the edge of the beach, face down in the seawater shallows that he'd come to love so much.[49]

In the end, Charles Darwin's theory of coral subsidence was proved substantially correct, although it took a combination of technical advances in drilling and American war-chest money to demonstrate it. During the 1950s, after American scientists had exploded a series of nuclear tests equivalent to seven thousand hydrogen bombs at Enewetak Atoll in the Marshall Islands, the U.S. Geological Survey set up two high-powered drills to cut through the atoll reef to investigate its resilience.

Although the deep cores did look like "old marine limestone," as Alex Agassiz had called it, they proved to contain coral fossils that could only have grown in shallow waters. Eventually, at 4,629 feet, the drillers reached a base of olivine basaltic rock that had been pushed up from the bowels of the planet. They concluded that, in accordance with Darwin's theory, Enewetak's coral reefs had begun to grow during the Eocene epoch, and for thirty million years or more had continued to scramble upward on a sinking volcano, thickening as the lava subsided. During that time, as Daly argued, the sea level had fallen and risen with the ice ages, the water temperature changing accordingly. This in turn helped to explain the different appearance of the crust of coralline limestone on the surface, compared with the much older platform beneath it.[50]

So the young Charles Darwin had been right all along, and in different ways both Alex Agassiz and Alfred Mayor were too blinded by their personal obsessions to see it. Still, the Sturm und Drang of their scientific arguments wasn't wasted. Because their quarrel thrust the Great Barrier Reef into the forefront of scientific attention, it fueled the study of coral reefs all over the world, initiating important new methods for investigating and understanding marine life. It was a legacy of which even squabbling Alex and Alfred might have been proud.

10

SYMBIOSIS

Cambridge Dons on a Coral Cay

THE PROGRAMS OF ACADEMIC SCIENTISTS are rarely treated as tabloid fare, but an announcement on September 3, 1927, by the British Association for the Advancement of Science proved an exception. To the surprise of many scholars, news that a group of young Cambridge scientists would live for twelve months on a coral island in Australia's Great Barrier Reef made headlines among Britain's mass newspapers. Perhaps the fact that Henry De Vere Stacpoole's steamy, bestselling Pacific novel *The Blue Lagoon* had just been shown as a silent movie gave the story some extra juice.

The expedition was also seen as romantic in other ways. The leader, Charles Maurice Yonge (always known as Maurice), a shy, rangy Cambridge don of twenty-seven with an endearing stammer, was to take his new Scottish bride, Mattie, twenty-four, to live with him in "the lonely coral wastes" of the South Pacific. Evening papers vied to describe the ordeals that lay ahead for the brave Mattie. She would supposedly be "the first English woman ever to live on the reef," spending the year "in a primitive hut on one of the low islands, with Queensland as the nearest civilized country."

Omitting to mention that she was a qualified doctor with an Edinburgh University science degree, the London *Evening News* noted that this "smiling happy girl barely out of her teens" was prepared to follow her husband anywhere and to laugh in the face of exile. The nearest shopping was a day's boat journey through shark-infested waters, but she would at least have a native "Man Friday" to help. Most versions of the syndicated story also left the purpose of the expedition until the last sentence—evidently they were going to study "the biological condition of the reef itself and the sea around."[1]

Serious newspapers recognized something odd about the expedition beyond its romance. It had come about at the suggestion of Australia's Great Barrier Reef Committee, made up of university and government scientists, and was to be half funded by Australian government money, yet it was to be led by a group of junior British marine scientists who'd never even seen a coral reef. Why were the Australians paying for such novices to study their Reef?[2]

Maurice Yonge, an expert on oysters though not on corals, explained that his group would be applying new biological and environmental methods developed in Britain and Europe to the study of growth patterns of corals and reefs in all four seasons. "Curiously enough," Yonge declared, "there are no Australians qualified for the task." British scientists would thus set up a temporary marine laboratory and research station for this purpose on the Low Isles, about forty miles from the mainland, where they would also investigate for the Australian government the commercial potential of Barrier Reef marine industries.[3]

The Cambridge expedition, as it became known, arrived in Australia in early July 1928, to be welcomed by state governors, politicians, and scientists at each major port on its procession from Melbourne to Cairns. Mattie Yonge was flattered that "our reception . . . was such that might have been given to royalty." Her husband, however, was warier: he'd been warned to prepare for a possibly skeptical reception from a prickly Australian popular press and scientific community. At every opportunity he took pains to stress that the project was as much Australian as British, pointing out to journalists that a young Australian economic biologist, Frank Moorhouse, had just been appointed by the Queensland government to study the cultivation possibilities of the Reef's marine resources, and would soon be joining them.[4]

As soon as the ship reached Sydney on July 4, Yonge and his deputy, Dr. Frederick Russell, met with the director and trustees of the Australian Museum to assure them that, provided funding could be found, they'd be pleased to host research visits from up to five members of the museum's scientific staff. Even so, that evening, at a dinner hosted by the local branch of the Royal Zoological Society, the Britons felt the frisson that their visit was causing among the museum's zoologists, who were proud of their own expertise in collecting and classifying Australian marine species.[5]

Talented, enthusiastic, and often self-taught, the museum scientists were a boisterous group. They included men like the American-born actor, dancer, and crab collector Charles Melbourne Ward, the self-proclaimed joker and ichthyologist Gilbert Whitley, and the rowdy zoologist's clerk Frank McNeill, who'd distinguished himself at Gallipoli for both courage and insubordination.[6]

When questioned about the absence of taxonomic zoologists in their ranks, Yonge conceded that such work was important and would be undertaken by Australian experts, but, within the specific context of the Cambridge expedition, it was necessarily "a side issue." British marine scientists, he explained, "were turning more and more from systematic to experimental zoology," which would be the main focus of their work on the Low Isles.

As payback for such donnish superiority, the fresh-faced young Britons then had to endure a succession of jocular speeches from the Australians, listing the perils that lay in wait for them on the Reef. These ranged from sharks, groupers, and giant clams to sandflies, mosquitoes, malaria, and sunburn. Next morning the tabloid *Sun* relished "Welcoming the Expedition" with the opening words: "Tiger-sharks, with great triangular teeth, slipping silently along; stone-fishes, with 13 poison-spines along the back, looking exactly like the rocks which they lie in waiting to be trodden upon."[7]

Sharks and stonefish were the least of Maurice Yonge's worries. As a biologist briefed to investigate the mysteries of symbiosis in reef-growing corals, he was aware that his expedition was a fragile coalition. It had been tacked together by two academic titans situated on opposite sides of the globe, who held opposing ideas about the priorities and methods of reef science.

Henry Caselli Richards, a professor of geology at the University of

Queensland, who'd initiated the idea of inviting British biologists to undertake the expedition, had spent the previous decade doing his best to sideline biological studies of the Reef. It was thanks to the genial charm and energy of this scientific Machiavelli that the Great Barrier Reef Committee (GBRC) had been founded in 1922, in collaboration with the Queensland government, and its mandate was to focus on the Reef's geological origins. A networker with extensive academic, business, and government connections, Richards's ambition had been to insert himself and his geology colleagues at the forefront of international debates over the validity of Darwin's theory of reef origins. More specifically, he hoped to make himself famous by proving Darwin right and contrarians like Alex Agassiz and John Murray wrong.

To this end he'd circulated an influential paper, "The Problems of the Great Barrier Reef" (1922), which favored Darwin's interpretation and cited six future scientific imperatives in order of priority. Geological surveys and reef drillings were listed first; biological experiments on corals came sixth. As chairman of the GBRC, Richards had also assiduously blocked the efforts of some members—including its secretary, mollusc scientist Charles Hedley—to establish a marine biological research station on the Reef. Instead, Richards diverted the bulk of the committee's money into an expedition to drill Oyster Cay, twenty-five miles northeast of Cairns. When the popular Hedley threatened to resign over this colossal waste of money, Richards somehow persuaded him to take charge of the drilling.[8]

The Oyster Cay expedition of 1926 proved as farcical as Hedley feared. Until the drill failed at 580 feet, it brought up only loose coral and sand. Bedrock was nowhere to be seen. Though Richards managed to fob off a threatened backlash from frustrated committee members, he soon found himself experiencing pressure from less pliable quarters. At the Third Pan-Pacific Science Congress in Tokyo that year, the doyens of international reef science were preoccupied with biological and ecological, not geological, issues. The final plenary session declared that "coral reefs are symbiotic entities whose origin and growth relations have received too little attention," and resolved to promote a "comprehensive investigation of the coral reefs of the Pacific."[9]

Sniffing the wind, Richards decided he should change direction and lead the charge to mount a biological expedition in Australia. But he was chagrined to find that the climate of discouragement he'd so carefully fostered had deterred Australian university graduates from working on marine biology. A search of the universities produced nobody. Hedley's death from an asthma attack on the way home from the congress further thinned the biologists' numbers. Still, this did provide a convenient scapegoat: Richards unhesitatingly blamed his lieutenant for having spent too much time working on Oyster Cay when he should have been undertaking a biological survey of the Reef.[10]

Hedley, however, had some posthumous vengeance when a journalist friend of his, S. Elliott Napier, had pungent things to say about the neglect of biological studies of the Reef. Napier's critique was published in a series of newspaper articles in December 1927, which arose from a "nature-study and holiday" expedition he had just undertaken to Bunker and Capricorn islands at the southern end of the Reef. His subsequent book incorporating the articles, *On the Barrier Reef*, was a bestseller, and reprinted every year for the next decade. In it he berated Australian scientists for having failed to study the sublime treasure that lay at their doorstep. "As a field for investigation," he thundered, ". . . the Great Barrier Reef affords the rarest of opportunities. The marine fauna, in particular, though in richness unequaled anywhere on the globe, has been but poorly studied." The outside world, he suggested, valued the Barrier Reef more than Australians did.[11]

Richards, the target of Napier's spleen, thought he'd better do something about this vacuum of Reef biologists. With the help of an energetic ex-governor of Queensland, Sir Matthew Nathan, he appealed to Britain for help in providing biology graduates and in raising additional funding for a proposed Barrier Reef expedition. To succeed, though, Richards needed to win over the acerbic titan of British marine science, J. Stanley Gardiner, professor of Zoology at Cambridge and head of the country's Fisheries Department.

Gardiner shared Richards's enthusiasm for marine economic development, but little else. He'd built his fame on biological investigations of marine flora and fauna among the reefs and islands of the Indian Ocean, where he'd become an expert on the physiology of coral nutrition and growth. On the question of reef origins, however, he was a fierce anti-Darwinist and thought all efforts to prove subsidence by drilling to be imbecilic. At a meeting of the Royal Geographical Society in 1925, he'd ventured the blunt opinion that "It is the greatest pity in the world [that] there is a Great Barrier Reef. Its existence is really a tragedy so far as the people of Queensland are concerned. It is a great nuisance to navigation. It is also a curse because it destroys 70,000 to 80,000 square miles of most admirable trawling ground."[12]

Even so, Gardiner knew a bargain when he saw one. Australia's offer to co-fund a twelve-month biological investigation of a Barrier Reef coral island offered an irresistible chance for British marine science to solve what he believed to be marine biology's foremost mystery: What was the exact role of the minute algae (zooxanthellae) known to be living within the cells of tropical reef-growing corals? Betting on youth, Gardiner engineered a lucrative Balfour Fellowship at Cambridge for Maurice Yonge, a bright young oyster physiology graduate from Edinburgh University, whom he then persuaded to lead the expedition.[13]

This was a stroke of organizational genius. A ferocious worker with a quiet charm, Yonge had served as a second lieutenant in the First World War, where he'd proven himself an excellent leader. Furthermore, though an expert marine physiologist, he was willing to learn experimental methods at Cambridge for a full term before being dispatched to Australia. Gardiner encouraged his new protégé to recruit like-minded young biologists—women included—who were likely to enjoy the adventure of working rough on a coral island for little money. In doing so, Yonge presumably took the advice of one Cambridge academic who told him, "Everyone thinks this expedition will be a very big thing, so don't invite any rotter."[14]

Rotters aside, both Gardiner and Yonge knew they were taking a big risk sending an expedition of young British scientific "experimenters" to a country renowned for its prickly nationalism, its suspicion of intellectuals, and active neglect of marine biological science.

On July 16, 1928, Maurice Yonge's cryptic daily notes reported that the Cambridge scientific party departed the Queensland port of Cairns for the Low Isles at noon, sailing in perfect weather and a calm sea on their motor launch, the *Daintree*. At 3:00 p.m. they sighted their home for the next year. It was low tide and a large area of fringing reef lay exposed. Adjacent to the flat stood "a circular mound of sand about 250 yd. diameter, surmounted by [a] lighthouse in centre and crescent-shaped mangrove about half a mile to west." Mattie Yonge thought this "one of the most beautiful sights I think I have ever seen." Two days later Maurice recorded the arrival of an expected visitor, a Mr. Charles Barrett of the Melbourne *Herald*.[15]

It's unclear who thought to embed a sympathetic journalist on the conjoined islands during the expedition's early weeks, and then to syndicate the resulting articles to newspapers in Australia's key cities, but it has the hallmarks of Richards's genius for publicity. The decision to select Barrett for the job was also inspired. A knowledgable nature journalist with several respected books to his credit, Barrett combined the romantic passion of his late friend Ted Banfield with the pipe-smoking bonhomie of a favorite uncle. The content of his seven subsequent feature articles show that he'd been carefully briefed. Even so, he faced a daunting task to translate the theories and experimental procedures of laboratory-inclined biologists into mass-media fare.[16]

In retrospect we can see that Barrett's approach to the task anticipated those modern TV reality shows that lure volunteers to undergo the rigors of running an Edwardian country house, or sailing an eighteenth-century coal bark, although in this instance the people, places, and challenges under scrutiny were genuine. Barrett wanted to convey to his audience what it felt like to be living and working with these young scientists on a "three acre flake of rock and sand." His narrative combined accounts of the dangers, hardships, and beauties of a coral isle with everyday details of human social life and with evocations of the wonders of biological research.

When the journalist landed on the island, the young scientists were still settling into their new home and preparing a work environment. Barrett cleverly turned these mundane tasks into a romantic saga of building a "village" and research station on a remote desert isle. Echoing Yonge's view that "a coral reef from the standpoint of a biologist is a particular community of animals and plants with a definite 'organization' of its own," he represented the expedition as an intricate working democracy. Men and women, junior and senior, white and black: all threw themselves unreservedly into cooperative tasks, both hard and humble.[17]

This was "science in shirt sleeves," and in blouses, khaki shorts, and sandshoes. "Youth is at the prow, but pleasure has not the helm," Barrett quipped. Maurice Yonge sawed planks to make bookshelves. Mattie dressed somebody's coral-scratched leg. Sheina Marshall, the phytoplankton expert, used a plane and hammer to improvise laboratory stools from packing cases. Other women heaved brimming buckets of seawater from the beach to the aquarium. "We are not ornamental, we've come here to work," they said.

Gender equity cut both ways: "men whose business it is to study plankton and echinoderms" were busy washing socks. Andy Dabah, the Aboriginal handyman from the Yarrabah mission, climbed up a coconut tree to fix the wireless antenna and tossed down half a dozen fresh nuts for the scientists. His wife, Gracie, was cooking for the party, while her two young children helped in the kitchen, or played on the coral strand with empty pickle jars.[18]

In no time they'd created a sophisticated laboratory, crammed with benches, high stools, glass vials, microscopes, a centrifuge, dissecting baths, an aquarium of colored fish, and rows of reference books. In the evening, tired figures wandered into their rough huts of Australian silky oak to rest briefly on camp beds, before gathering in the long room that doubled as the laboratory and communal dining room. The facilities had cost a mere six hundred pounds, Barrett pointed out admiringly, and he estimated the whole expedition's costs at ten thousand pounds (it proved considerably cheaper).

There was also just enough time before dinner to contemplate "the end of a perfect tropic day and the night is still and full of stars, with the sea quiet and no sound from the outside world but the cry of a homing tern. The coconut palms print their fronds in the shadow on the coral sand, as the lighthouse beams go sweeping by, making long silver pathways for miles across the water."

Afterward the little group of scientists gathered round hurricane lamps to discuss everything from the behavior of bêche-de-mer to the latest novels. Some played bridge; others wrote letters, read books, or prepared to go fishing in the "flattie."

"Do you like it here?" Barrett asked a couple of the women. Gweneth Russell responded for them all: "I think it is wonderful." Barrett was not surprised: "waking early, they hear honey birds calling in the palms and waders crying plaintively along the coral strand." What a way to start the day.

Yet, Barrett reminded his readers, "a scientific expedition has serious work to do." The shore party, led by Thomas Alan Stephenson, assisted by his wife, Anne, was studying reef animal ecologies. They aimed to isolate "typical areas" within different environmental zones. They'd already mapped out a ten-foot-square "coral garden," photographed it in sections, and counted its inhabitants. They were also dredging up samples of seabed mud and sand for examination in the lab. Their eventual aim was to work out the breeding habits and seasonal behaviors of sixteen key marine animals, "about which little is known." These included giant clams, sea urchins, corals, crabs, and bristle worms. Many species appeared to show some strange connection between breeding times and particular tidal and lunar phases—another mystery to be unlocked.[19]

Working with the shore party was a "physiological group" of Maurice, Mattie, and a young Australian student, Aubrey Nicholls. They aimed to investigate "the family life of corals," especially how corals feed and grow—what Maurice thought "the most fascinating [problem] marine biology has to offer." Among other experiments, they used clamshells, drainpipes, and logs to attract coral "settlers," and constructed bases on which to graft varieties of coral species. After distributing these bases over a range of reef environments, the scientists would monitor the differing growth rates of the new coral. This was vital information because reefs were always competing with the forces that worked toward their destruction, which ranged from burrowing crustaceans to coral-eating fish and torrential rainstorms.[20]

The Australian Frank Moorhouse was also one of the shore party, and Barrett observed that his work as an economic biologist would have vital significance for the future wealth of Queensland. The marine creatures Moorhouse was investigating included smoked bêche-de-mer for the Chinese market, and the crimson- and red-banded trochus shells whose nacreous interiors were used in the Japanese button industry. Moorhouse had found a little "nook" in the fringing reef from which he could observe the live snails of the trochus, which were also easily gathered by hand at low tide. Andy Dabah would later build Moorhouse separate trochus and oyster pens for his experiments, but prior to that, Barrett reported, Moorhouse worked in the lab with a microscope, "seeing what no one had ever seen before—the very beginning of *trochus* shell life."[21]

Barrett also wrote about going plankton hunting with the "boat party" in the waters surrounding the Low Isles, which he claimed resulted in some of his strangest experiences. Frederick Russell, Gweneth's husband, stalked zooplankton, the microscopic animals that float almost unseen in every drop of reef seawater, and on which scores of predators feed. Sheina Marshall chased phytoplankton, especially the tiny vegetable diatoms that make up the "pastures of the sea," and which are the base of a food chain that works up through corals and fish to humans. A. P. Orr, the chemist, was investigating the chemical compositions of plankton environments, testing samples of seawater for salinity, phosphates, nitrates, oxygen, and organic materials. He was also measuring the light densities of seawater—something Barrett joined him in—because phytoplankton, being plants, were dependent on light for energy.

Barrett regarded plankton catching at different water depths as very like big-game hunting, except that it was a "scientific sport . . . with a definite object [and] its results are of economic value." In these seawater safaris, "the score is made up of problems solved, secrets of Nature's ways discovered." More mysteries lurked unexplored in the waters of the Low Isles, he claimed, than in the jungles of New Guinea. But instead of guns, the researchers' weapons were fine-meshed trawling nets, insulated water bottles, depth recorders, and tin cylinders painted black so as not to attract sharks. Barrett felt a frisson of anticipation as he peered into the collecting jar, hoping that some new species of pelagic life would be wriggling among the mass of copepods, crab larvae, tiny jellyfish, Sagittoidea (arrow worms), diatoms, and Foraminifera (minute, single-cell protozoans). He was also excited because he'd only seen such bizarre creatures in drawings: "Believe me it is revelation to spend even a day in a tropic sea, with plankton hunters, who hunt for research."[22]

Some aspects of Low Isles research also carried the thrill of real danger. The expedition was equipped with a specially built diving helmet modeled on one pioneered by the American biologist Alfred Mayor. Made of galvanized iron and with two glass windows, its air supply depended on continual hand pumping. Frederick Russell, looking like "Ned Kelly preparing for a police offensive," was the first of the expedition's scientists to use it, observing the radiant corals below the reef's edge.

Barrett was envious until he tried a shallow dive himself, to be abruptly reminded of wearing a wartime gas mask. And down there, he couldn't help thinking about man-eating tiger sharks, groupers with indiscriminate jaws, giant clams (*Tridacna*) waiting to clamp on the unwary diver's foot, or the toxic fangs of sea snakes, just like the six-footer he'd seen over near the mangroves. Still, these young scientists seemed as brave as they were clever: "nobody is worrying at all about the perils of the reef," Barrett wrote. Sensibly, they were more concerned about avoiding everyday annoyances like the sting of jellyfish, the scratch of jagged corals, and the needle-sharp spines of sea urchins.[23]

But the peril was real enough the day Barrett traveled with the boat party to Batt Reef, on the outer Barrier, where they hoped to view "a section of the submarine structure which ranks as one of the wonders of the world." A stiff breeze made the boat "curtsy" as they passed beyond the shelter of the mangroves. This turned into "a proper dusting" as they crossed the eight miles of open water to the outer reef. Eventually they sighted a line of breakers and the shadows of "ominous brown and yellow 'isles' beneath the surface." Soon they were caught in a coral maze. Quickly they posted a lookout at the masthead and others at the prow. Despite these and a detailed chart, the helmsman had "an anxious time" as they missed a patch of submerged reef by less than half a boat length.

Barrett was reminded of Cook's incredible navigating ordeals among these "coral strewn waters," an empathy that grew as they were battered by "swinging seas" on the return voyage: "one huge wave crashed aboard the *Luana*, and all the way home to our island she was dodging Pacific rollers large enough to make a coastal steamer shudder."[24]

Barrett's final article ruminated about symbiotic cooperation on the Reef. Despite the ferocity of the struggle among most reef species, some marine creatures chose to become what he called "mess-mates"—not unlike the Low Isles scientists themselves. These mess-mates, however, could display radically different styles and degrees of friendship. Some species, parasite-like, exploited their hosts; others were commensal, with one species prospering from the partnership at no cost to the other; and some were mutually cooperative, with the relationship benefiting both partners. Maurice Yonge's chief task on the expedition was to find out which of these modes fitted the puzzling association between reef-growing corals and the tiny brown algae living in their tissues.

Yonge had found a range of examples of mutually cooperative relationships among the marine creatures of the Low Isles reefs to use for comparison. Barrett himself had captured a small shrimp—"a lively red-tailed midget"—which lived in the body cavity of a giant anemone. Another, smaller anemone gave refuge to a scarlet-and-white-banded pygmy fish. One species of bêche-de-mer even endured what we might regard as the indignity of having a glass eel take up tenancy in its anus. Gall crabs built a home within the rocky cavities of the reef itself, influencing the madreporic corals around them to form a small living room for the female, complete with a tiny passage to the water. Such was love on the Reef.

As he prepared to leave the Low Isles, Barrett confessed to finding this marine research endlessly fascinating, and he envied the young scientists their future year there. They were happy people, both because they were working "in a fairyland" and because, as Alfred Mayor had written, "love, not logic, impels the naturalist to his work."[25]

Charles Barrett's newspaper articles constituted a remarkable piece of journalism, much of whose accuracy was confirmed in the details of Maurice Yonge's prosaic daily notes. Barrett's syndicated articles generated so much local interest that crowds of curious picnic parties began appearing on the Low Isles, forcing the scientists to fence off their experimental areas and equipment. Still, they couldn't complain: Barrett had won for the expedition the nationwide enthusiasm of Australia's public and scientific communities.[26]

Of course, he'd sometimes idealized or oversimplified the story. Some omissions were a result of his early departure. He never saw, for example, Yonge's occasional flares of exasperation, such as when the Australian Museum conchologist and ichthyologist Tom Iredale brought his young son and a guest to stay on the island without prior consultation. He didn't see the irritating amount of time Yonge had to spend managing finances, entertaining visitors, organizing work rosters, and conducting meetings to report on research progress. And the sympathetic journalist didn't stay with the expedition long enough to observe the sheer physical and mental exhaustion entailed in working long hours in tropical humidity, heat, and rain. "Let no one think that life on a coral island is unending bliss," Yonge later wrote ruefully.

> Our work was never finished. All day, and not infrequently, all or part of the night, it continued. Work of such intensity in such a climate was hard; I felt perpetually tired; every action demanded a tremendous

initial effort . . . It was not until I left the island that I realized fully under what a strain we had been living for many months past. At the same
time it was work that made life endurable: without that compelling interest the continuous association of so many people in such an environment would have been impossible.[27]

Neither was the situation of the Aboriginal workers quite as idyllic as
it had initially seemed. Gracie proved to have a quick temper, and she
chose to leave the island after four months. Andy, in the meantime, had
apparently done little but stand motionless, staring at the water with fish
spear raised. Yonge probably didn't notice that he was unwell—he died
within a few months of departing the island.

However, the replacements from Yarrabah mission, Minnie and Claude
Connolly, and their children Teresa and Stanley, were a greater success.
Minnie did all the cooking and washing without complaint. Claude,
though elderly and lame, delighted everyone. A former tracker who'd
been wounded in a shoot-out with the Kelly gang in Victoria, he never
hesitated to take on any challenge. When not working, he carved gifts
for the scientists of firesticks, spears, and model canoes, and he loved
polishing pearl shells on a grinding wheel to give to them. Maurice
Yonge thought him "an honour to his race" and "a living refutation of the
slanders from which the Australian aboriginal has so long suffered."

Yonge also praised the expert contributions of the two Yarrabah boat-
men, Harry Mossman and Paul Sexton—"good servants to the expedition . . . and to science." Being paid twenty-two shillings and sixpence a
month plus free tobacco, at a time when equivalent white male workers
earned twenty pounds a month, they obviously weren't in it for the
money.[28]

Barrett's early departure from the Low Isles also meant that he missed
Yonge's coral physiology experiments, which were designed to investigate the feeding, digestion, excretion, and respiration of reef-growing
corals, and especially their relationship with symbiotic algae (zooxanthellae). A few coral experts, including Stanley Gardiner, speculated
without real proof that the tiny brown algae played an important role in
coral nutrition. Others thought that the plants might boost the corals'
energy by supplying oxygen. Yonge was skeptical on both counts. He
suspected that the relationship was at best commensal, with the algae

the sole beneficiaries, or possibly even parasitic, at the expense of the coral hosts.[29]

It was obvious how algae benefited from the association. Insinuating themselves into the very cells of the coral polyps, they were thus protected from predators while still afforded access to light through the thin tissues of the coral. Like all plants, the algae needed sunlight to create energy through photosynthesis, and in the process they produced oxygen. Using their light-activated coloring matter chlorophyll, the algae built up their basic organic food of starch and sugars from a mixture of seawater elements produced by the corals. They "obtain from the excretion of the corals valuable food material," Yonge wrote, listing these nutritional wastes as carbon dioxide, ammonia, and hydrogen. With both protection and their food and energy on tap, the little plants could hardly have found a more perfect place to live.[30]

It didn't necessarily follow, though, Yonge thought, that the corals received reciprocal benefits. Such an assumption was sentimental rather than scientific, a throwback to the socialistic beliefs of pioneering biologists like Patrick Geddes and Prince Peter Kropotkin, who'd believed that cooperation and not war was nature's way. To test this cooperative proposition with proper rigor, Yonge designed a series of elegant and trenchant experiments. He began by showing that corals were wholly carnivorous, and greedily so: they gobbled up zooplankton and shunned all vegetable enticements. And though Yonge thought these coral polyps might indeed derive some energy from oxygen, he didn't see why they couldn't supply this themselves by processing seawater. His experiments also showed that during daylight the algae produced far more oxygen than the corals needed: their guests' supply was thus surplus to requirements.[31]

Two further experiments appeared to reveal that coral polyps did not rely on their algae tenants for survival but continued to live in the dark, apparently by eating zooplankton. When polyps were deprived of zooplankton they became markedly distressed, but well-fed corals survived whether they were in light or dark environments. Yonge did notice that after five months in the dark, corals lost all their color, but he saw no evidence that this bleaching did them any real harm, for they "were otherwise in perfect condition." It did not occur to him that bleached corals might no longer be capable of building reefs, or that they would in fact eventually die.

Ultimately, Yonge was prepared to make two partial and tentative concessions to the idea of a cooperative symbiosis between corals and algae: "If we must have a 'use' for the plants, then I think that the speed with which they dispose of the waste products of the corals increases the efficiency of the latter, while they [the algae] certainly provide abundant supplies of oxygen, without which it is just possible that such immense aggregations of living matter, which constitute the coral reef . . . could not originate and flourish."[32]

Maurice Yonge's half-concessions to the cooperative role of zooxanthellae were later proved correct, but he failed to realize the degree to which reef-growing corals are dependent on their algae partners for the oxygen-based nutrition and energy needed to build reefs. His experiments also seemed so original and persuasive that they went largely unquestioned for the next thirty years. Still, though flawed, his findings would provide the baseline for an explosion of new work on coral symbiosis by scientists in the West Indies, the United States, and Europe during the 1960s and 1970s.[33]

Today we know that because of their symbiotic relationship with algae, reef-growing corals are autotrophic, meaning they are "predominantly self-sufficient in supplying nutrition from their own biological processes." Although polyps do graze on small amounts of floating plankton, the available supply doesn't nearly meet their needs as reef builders. Reef corals rely on the solar panels of their algae tenants to provide the prodigious energy required to produce lime-based skeletons, in the same way that we make bone. Geneticist Steve Jones points out that such algae-assisted corals are three times as efficient in the light as in the dark, and in perfect conditions can have their productivity increased a hundredfold by their tiny helpers. Thanks to the zooxanthellae waste, reef corals are able to lay down carbon at nearly twice the rate of a rain forest, "making shallow water coral reefs the most productive natural places on the planet."[34]

There are many corals that are not algae-assisted, but all reef-growing corals are. They need the extra energy generated by the algae's oxygen and sugars to grow fast enough to combat all the forces that work toward reef destruction. Maurice Yonge did not quite get this point, but his work was essential in allowing those who followed to make the discovery.

Maurice Yonge was right, though, not to idealize the relationship of polyp and plant. Their partnership is no cooperative utopia. Each partner pays a price for the contract, which seems to be built on nothing but tough Hobbesian self-interest. Though their symbiosis has survived for something like 240 million years, and has produced the Great Barrier Reef in the process, it can, and probably will, break down one day. As we'll see in later chapters, if forced to go it alone for the sake of self-interest, corals and algae will part. Each might perhaps survive alone, but they will no longer build reefs. Steve Jones warns us: "In fact the submarine union is always on the edge, with the guest in constant danger of forced expulsion or voluntary exile, and its host of a solitary existence that may be bad for its health. The arrangement thrives when times are good, but may split up when they get nasty."[35]

The Cambridge expedition's body of scientific findings was presented in seven massive volumes between 1930 and 1968, as well as in scores of papers and books. Its global impact on coral reef science and tropical marine ecology can hardly be exaggerated. There'd been nothing like it before, and there have been few more influential expeditions since. Low Isles data is still admired and used today.

Within Australia, one emblem of the success of the partnership was the decision by the British Reef Committee to donate the Low Isles buildings to the Queensland government for a permanent marine research station. It was to be led by Frank Moorhouse and to focus on economic biology, but sadly it didn't last long. A cyclone leveled the buildings in 1934 and this, coupled with the indifference of the Queensland government, pushed Moorhouse to resign a year later. After that the station petered out.

A more enduring by-product of the expedition was the inspiration it gave female scientists everywhere. Six of the fourteen members of the biological section were women, an astonishing statistic for its day. Sidnie Manton joined the expedition at a later date: she had received top marks in zoology at Cambridge four years earlier as a twenty-two-year-old in Girton College, but was not permitted to take the prize for being first in her class because of her gender. Already famed for her work on crustacean embryology, she threw herself into a quantitative ecological survey of Low Isles marine life, which, according to Yonge, achieved more in four months than the rest of the party over the full year. Manton later became one of the first women fellows of the Royal Society. For her part, Sheina Marshall ended her stellar career as deputy director of the University of London's famous marine research station at Millport, Scotland.[36]

News of their colleagues' presence on the island acted as a magnet for Australian women scientists. Dr. Gwynneth Buchanan, a lecturer in the University of Melbourne's zoology department, joined them for a short visit, and a botanist, Mary P. Glynne, made a brief survey of the Low Isles in April 1929. Perhaps the most illustrious visitor was Freda Bage, biologist and head of the Women's College at the University of Queensland. When the Great Barrier Reef Committee learned of the forthcoming contingent of women from Cambridge, they quickly moved to appoint Freda Bage as their first female member. She hosted all the expedition's women scientists in Brisbane, organized talks to graduates, and energetically publicized their work.[37]

The expedition's contribution to the economic biology of the Reef proved less impressive. Moorhouse's work on the life histories of trochus and goldlip oysters was genuinely original, but Yonge's own "Economic Report" was lame by comparison. Although the latter was quoted approvingly in local Australian newspapers, Yonge's findings had been cobbled together during a short trip to the Torres Strait on his way back to Britain. Most of his recommendations were either contradictory or perfunctory. Despite noting a serious depletion of green turtle stocks, he concluded that this species, along with dugong, bêche-de-mer, and hawksbill turtle, could be more efficiently exploited through improved scientific methods. At the same time he urged the Queensland government to impose tougher restrictions on hunting dugong, and to develop a plan for the biological study of all potentially commercial species, including sardines, which were abundant in the waters of Murray Island.[38]

Maurice Yonge's most important contribution to the Reef's future economy was partly unintentional. His popular memoir, *A Year on the Great Barrier Reef* (1930), projected a time "not far distant when . . . the sand cays of the Capricorns will each possess its hotel and guest house." The modest Maurice never guessed he would be an instrument of his own prophecy, and that the Cambridge expedition and his book would ignite new fascinations with the Reef, among tourists as well as scientists, on both sides of the world. No scientist had written about the Reef with such a combination of passion and precision since the neglected geologist Joseph Beete Jukes, nearly a century earlier.[39]

Yonge recalled that one of his most enduring memories was the sight and sound of Pacific breakers smashing onto the limestone ramparts of the outer Barrier Reef.

> Towering far above, 15 or 10 feet high, they came curling over and hung poised for a moment of time, grey and green and lowering under a dull, storm-racked sky, to fall with the noise of thunder and in a cascade of foam that came rushing up, a swirling flood to the summit of the reef. One was in the presence of forces far beyond the control, almost beyond the conception of man . . . The reef platform, now awash, was the only "land" to break the surface of the sea, and that land . . . with infinite labour built up by the lowliest animals and plants, adding grain by grain—nay, atom by atom—to their skeletons of lime, uniting, spreading, consolidating, resisting the fury of the sea and attacks of enemies, never resting, nor yielding, but converting the very substance of the sea itself into a submarine mountain chain of limestone, against which even the tempest-driven Pacific is utterly unavailing.[40]

One of Yonge's admiring Australian readers was a schoolteacher named Edwin Montague Embury, from the northern New South Wales country town of Manilla. The son of a wool scourer, "Mont" had grown up an ardent nature lover, before enlisting with the Australian Imperial Force in 1914. He returned to work in a bush school after the war, and there he "taught himself the essential elements of Reef and Coral structure and the creatures which inhabit this unique environment." His accurate and delightful little school primer, drawing on the writings of Banfield, Saville-Kent, and Yonge, was one result.[41]

In 1927 Mont joined a small "nature study and recreation" outing to Bunker and Capricorn islands, organized by E. F. Pollock, a councillor of the Royal Zoological Society of New South Wales. Inspired by the experience, Mont decided to stage a series of larger "Scientific expeditions" to the Whitsunday Islands, which would combine learning and recreation for around a hundred people, mainly schoolteachers. While exploring the

Whitsunday Passage on his first expedition, he'd "discovered" Hayman, "a beautiful primeval little island, with trees, birds and beautiful coral." In December 1932, after mounting eight further expeditions in the Whitsundays, he opened "The Hayman Island Biological Station."[42]

Modeled on Yonge's Low Isles station, it consisted of rough huts supplemented by tents, and a large central hall for dining, dancing, lectures, and research. An Embury relative, "Uncle Tom," acted as chef; two motorboats provided excursions for fishing and scientific dredging; and a shark-proof enclosure, a crude tennis court, and several sets of homemade underwater goggles offered recreational variety. Of course, at one level it was simply a tourist resort. Some regarded Embury's scientific pretensions as a ruse for making money under the nose of the Great Barrier Reef Committee, which restricted tourist developments. And there's little doubt that Mont nursed commercial ambitions, for he also took out provisional leases on a series of other islands.[43]

Even so, the scientific dimensions of the expeditions seem to have been genuine enough. Embury trumpeted his links with Yonge's expedition in his "Biological Station" advertisements, boasting that three of his leading lecturers, Frank McNeill, Bill Boardman, and Arthur Livingstone, were former Australian Museum members of "the British Low Isles Expedition." Three others, Mel Ward, Harold Fletcher, and Joyce Allan, were currently working with the museum in a collecting capacity. Moreover, the administrators of the museum, though wary at first, did eventually begin to advertise and exhibit the marine collections brought back from Mont's expeditions.

Paying visitors to Mont's Hayman Island biological station were offered lunchtime and evening lectures accompanied by magic lantern slides, on such subjects as "The Form and Character of the Great Barrier Reef," "Coral Reef Animals," "Sea Stars and their Allies," and "An Outline of Animal Classification." Other scientific activities included participation in a Hayman Island bird census, and the tagging of a thousand birds under the supervision of a leading ornithologist, William MacGillivray. Mont's brother, "Arch," designed and used one of the first Australian underwater cameras, with which he photographed Mel Ward "swimming around the bottom of a pool probing amongst the rocks with his prospector's pick." Several "Biological Station" organizers also became pioneers of Reef conservation. From the early 1930s Mont and Frank McNeill, in particular, campaigned against the unsustainable hunting of Torres

Edwin "Mont" Embury's 1920s "nature study and recreation"
expeditions to the Whitsundays were the forerunner of ecotourism.
(Mitchell Library, State Library of New South Wales)

Strait pigeons, green turtles, and dugong, and the profligate collecting of rare shells.[44]

Mont Embury's pioneering ecotourist Reef expeditions were hurt by the Great Depression of the 1930s and then terminated by the Second World War. His undeveloped island leases later became wildlife sanctuaries. Yet his idea of popular scientific tourism didn't die altogether. It was to be revived after the war by others, including Kitty and Noel Monkman, a remarkable pair of conservation-minded filmmakers and naturalists who still await their biographer.[45]

When the world-famous marine biologist Professor Maurice Yonge returned to the Low Isles for a holiday in 1972, he was distressed at the damage caused by pollution and uncontrolled tourism in the once lovely habitats. He was even said to have regretted his own role in publicizing the wonders of the Great Barrier Reef. Perhaps he might have felt less guilty if he'd known that his expedition and book had also inspired Australians who wanted to understand and conserve those same wonders. These disparate but passionate individuals would in a few years join together to save the Reef, in what was to prove a long and brutal battle.

11

WAR

A Poet, a Forester, and an Artist Join Forces

L IKE MOST WARS, it began with a skirmish. Midway through 1967 the
poet Judith Wright, who was also president of the five-year-old Wildlife Preservation Society of Queensland (WPSQ), received a note. It was
from John Busst, leader of a tiny new branch of the society in Innisfail,
a coastal cane town one thousand miles north of Brisbane. Busst had
just read in a local paper that a Cairns sugarcane farmer had applied to
the Queensland government to dredge and mine Ellison Reef, an isolated clump of coral near Dunk Island. Claiming that the reef was dead,
the farmer wanted to mine its limestone to use as cheap fertilizer on his
canefields. Busst, sensing "a vital test case," had immediately lodged an
objection. Would the foundation branch of the WPSQ in Brisbane be
willing to do the same? Judith Wright, already campaigning to protect a
variety of Queensland habitats, quickly complied.[1]

She knew John only by repute. Len Webb, a Commonwealth government forestry scientist and the society's vice president, had helped Busst
to found the new branch the previous year, recommending him as a long-
time friend and a potent asset to their fledgling cause. Busst belonged to
a post-Banfield tradition of dropout artists who'd left the rat race for un-
buttoned lives among the Barrier Reef's tropical islands. But he was
also, Len stressed, a highly educated, energetic, and practical man, with

private money, influential connections, and a genuine love of the rain forests, reefs, and waters that abutted his house.

People often found John Busst hard to read. His neat dress, precise movements, and trimmed gray mustache struck the Reef writer Patricia Clare as more typical of a colonial administrator than a bohemian artist. Len Webb admired his friend's practicality above all. A decade earlier, John and his elegant wife, Alison, had moved from Bedarra Island, in the Family group, to Bingil Bay, on the mainland near Innisfail. Together they'd built a white-and-blue-trim bungalow with wide verandas, hand-made bamboo furniture, and sweeping views of the sea. In typical fashion John had also underpinned it with a fortresslike concrete substructure to resist the ferocious local cyclones.[2]

When Judith Wright eventually did meet Busst, she was struck most by his bonhomie, finding him a "slender, enthusiastic man full of laughter, a compulsive smoker and a lover of good company." He had the gift of making friends from all walks of life, and of invigorating them all.[3]

John Busst threw all his energies into trying to stop the mining of Ellison Reef. He began his campaign with letters to local newspapers, arguing that there was no such thing as a "dead reef." He said he'd seen cyclone-battered reefs at Bedarra Island, where most of the corals had been smashed, yet they remained hubs of marine life. Len Webb had taught him, too, that all reefs were complex and interconnected biological communities that could never be treated in isolation. Currents and waves would carry the smothering silt from any specific mining site to corals and breeding habitats far and wide.

When John Busst tried to put these points at the first hearing of the case before the Innisfail local magistrate, in September 1967, he made no progress. His marked-up books and his accompanying marine expert—a local caravan park owner, Bill Hall, with years of practical knowledge—were ruled inadmissible. The magistrate told John that he ought to have brought properly credentialed scientific experts to testify in person. Though he managed to squeak an adjournment, John now needed scientific help urgently.[4]

But scientists proved elusive. His first attempts to recruit support from biologists at the University of Queensland and the Great Barrier Reef Committee drew a blank. The academics didn't want to involve themselves in local political squabbles, and anyway they saw no reason to dispute that Ellison Reef was "dead." The GBRC, being an advisory body to the Queensland government, had long been forced to accept the inevitability of the Reef's economic development. Biologist Robert Endean, the committee's president, hoped merely to confine economic intervention to partitioned sectors of the Reef; moreover, he thought Ellison to be an isolated, half-dead lump of coral with no distinctive biological significance.

John would have turned to Len Webb for scientific contacts and advice, but by ill luck the ecologist was in Europe on sabbatical. "Come ''ome,'" John begged by letter, ". . . I desire conversation with you—about 24 hours straight . . . will do! You got me into this, you bastard—and I'm enjoying every moment of it, so is Ali, who is even more ferocious (if possible) than I am!" He signed himself "The Enraged Amateur."[5]

Infuriatingly, only fellow amateurs seemed willing to testify, though Judith Wright did eventually find some trainee professionals. She introduced John to a group of graduate science students from the Queensland Littoral Society, led by a talented twenty-three-year-old zoologist at the University of Queensland called Eddie Hegerl. He and two other young divers immediately volunteered to make an underwater investigation of Ellison Reef, provided John could help to fund it and organize the logistics. Within a few weeks, he'd persuaded friends to donate free airfares, boat services, and diving-equipment hire for a five-day survey of the reef. The Bussts offered their own house as a base.

Soon after, John lobbied for help of a different kind from another amateur diver: Harold Holt, the Liberal prime minister of Australia, and Busst's oldest friend. They'd gone to school and university together, and John had later introduced Holt and his wife, Zara, to the pleasures of swimming, sailing, and spearfishing off Bedarra Island. Reveling in these sensual escapes, the Holts eventually bought a small holiday cottage at Bingil Bay, just around the corner from the Bussts. When John flew to Canberra to make his case, though, his old friend proved elusive. After five frustrating days of chasing him around Parliament House, John eventually caught Holt sleeping on a plane and forced him awake to hear a harangue about the threat to Ellison Reef.

It was not the only reef at risk, John insisted: the development mania of the Country Party–led Queensland government and its premier, Frank Nicklin, was limitless. Reliable rumor had it that the state government

wanted to use Ellison as a precedent for granting scores of mining ap-
plications for oil and gas—most involving the drilling or seismic blast-
ing of coral reefs.

Holt was genuinely sympathetic, but he was hemmed in by political
constraints. The push for states' rights, along the lines of the Queensland
government, was also a potent electoral force in federal politics. No
Commonwealth government had ever tested the legality of Queensland's
vociferous claim to own the Reef, but even so, Holt told Busst that if the
Ellison mining claim succeeded and the Reef was endangered, "I will
promise you personally that the federal government will take over the
Barrier Reef."[6]

At last Judith Wright found an impeccable scientific authority to testify against the mining of Ellison. Dr. Don McMichael was both an
eminent marine biologist and director of the Australian Conservation
Foundation (ACF), an elite body established in 1964 to provide expert,
objective, nonpolitical conservation advice to government and business.
The foundation was respectable, conservative, and a proponent of the
"rational exploitation" of the Reef, but it had a few fiery spirits like Judith Wright and Len Webb among its commissioners. By great good
luck, too, McMichael revealed that he'd surveyed Ellison Reef himself
in 1965.

When the Innisfail court resumed in November 1967, McMichael
explained to the government-appointed mining warden that Ellison, like
most reefs, contained both dead and living corals. Furthermore, he'd
found a species of mollusc there that was unique to Queensland waters.
This, when combined with Hegerl's report of having seen 190 species of
fish and eighty-eight species of live coral on the reef, was enough to persuade the warden to recommend against the mining application—a rul-
ing grudgingly endorsed by the minister for mines, Ronald Camm, six
months later.[7]

This small local victory would ultimately change the face of the conservation movement for decades, and the fate of the whole of the Great
Barrier Reef. It marked the moment when the Reef became the central
cause of conservationists all over Australia, and it unleashed a fourteen-
year campaign, "Save the Reef," that Judith Wright would later call "The
Coral Battleground" and John Busst "The Battle."

There'd been earlier ruckuses, of course. At the beginning of the
1960s the underwater filmmaker and amateur naturalist Noel Monk-

man, who lived with his wife, Kitty, on Green Island, became so worried about the depredation by coral and shell collectors and Taiwanese fishermen that he publicized the need for a Great Barrier Reef marine park. He was among those, too, who at Green Island in 1963 discovered the first coral damage wrought by crown-of-thorns starfish—an infestation of which was soon reported to be spreading all over the Reef. Monkman and several other naturalists were adamant that human behavior, such as overfishing and pollution, was responsible for the plague.[8]

Some Australians were also alerted to these dangers by a superb trilogy of books about America's coasts and seas, written by biologist Rachel Carson. Shortly before her death, Carson caused a further international sensation with the publication of her searing *Silent Spring* (1962), which exposed the pervasive spread of toxic chemicals into habitats worldwide. Needless to say, coral reefs were among these.[9]

In June 1967 Wright, Webb, and Busst had been made aware of specific problems on the Barrier Reef by a short article in the Wildlife Preservation Society's own magazine, written by their diver allies in the Littoral Society. After making the point that Queenslanders "hold the Barrier Reef in trust for future generations throughout the world," the authors argued that a Florida-style marine park should be established on the Reef to protect declining fish-breeding habitats, prevent increasing damage to water purity and food chains, and attract nature-loving tourists.[10]

Following the Ellison case, and the confirmation of the rumor that the Queensland government was opening most of the Reef to oil prospecting and other forms of mineral extraction, local attacks by conservation groups ballooned into full-scale war. Joh Bjelke-Petersen, who became premier of Queensland in August 1968, didn't care that the government had lost the first skirmish, or that a few eccentric intellectuals were outraged at the prospect of oil mining; he had the hide of a rhinoceros and the mind-set of a hyena. After instructing Minister Camm to reiterate the government's right to mine the Barrier Reef when and where it chose, he ordered him to have the entire length of the Reef zoned in preparation for leasing. Since Camm had already secretly issued oil-prospecting leases to six petroleum companies, even this proposed zoning was window dressing.[11]

Although Robert Endean, chairman of the Great Barrier Reef Committee, supported the government's oil policies, he was so angered by Camm's indifference to the starfish plague that he quietly leaked the details of these leases to the Wildlife Preservation Society. John Busst in turn passed on the news to the Queensland public. He told the local *Advocate* newspaper that the government had secretly completed a geophysical mining survey of the entire Reef province, an action that showed its willingness to jeopardize Australia's oceanic food supplies and to squander a natural asset whose "potential for research is inestimable and . . . aesthetic value for untold millions of tourists incalculable." If Bjelke-Petersen was not stopped, Busst warned, "we shall earn the unenviable reputation . . . as those Australian barbarians who destroyed one of the seven wonders of the world."[12]

Although the Ellison Reef campaign had forged an alliance among Wright, Webb, and Busst, the tiny trio looked ill-equipped to lead a war. A poet who'd barely visited the Reef, an obscure government forestry scientist, and a dropout artist seemed unlikely figures to halt the juggernaut of politicians, local developers, and international mining interests.

Judith Wright, born in 1915 in the New England region of New South Wales, grew up in an arresting landscape. A solitary child, she fell in love with "my blood's country," the lean, dry highlands around her family's ranch. Blessed with a brilliant imagination and a gift for language, she began from an early age to feel alienated from the conservative gentry values of her peers. Though she never stopped loving her family, she came to feel that they, along with most Australian settlers, had dispossessed the land's ancient original peoples and fostered a rapacious culture in their stead.

Judith was a romantic in the profoundest sense: someone who strove through the power of language, myth, and symbol to absorb the harsh beauty of the Australian continent and the environmental ethos of its Aboriginal peoples. Like the writer Patrick White, her distant cousin, she wanted to create "a country of the mind," instead of a country of mindless greed. This sense of dissonance between the Australian landscape and the culture of its colonizers pushed her into unorthodoxy. As her daughter recalled, Judith's sense of estrangement led this strong, intelligent woman always to side "genuinely and passionately with the outsider."[13]

Jack McKinney, a former drover and war-veteran-turned-philosopher who later became Judith's husband, helped give intellectual shape to this dissidence. Living together in an old timber cutter's cottage among the rain forests of Mount Tamborine on Queensland's Gold Coast, the couple were united in their dislike of modern Western civilization. They were repelled by the lust for money and progress, the brutal ideologies and technologies of power, and the indifference to humane moral values. After reading works like *Man and Nature* (1864) and *A Sand County Almanac* (1949) by American conservationists George Perkins Marsh and Aldo Leopold, they concluded that Australia's redemption must come through a revolutionary change of heart that embraced the Aborigines' ancient ethic of "caring for country."[14]

Friendships with two like-minded women, Aboriginal poet Kath Walker (Oodgeroo Noonuccal) and wildflower artist Kathleen McArthur, inspired Judith to join local conservationist campaigns to stop patches of rain forest being gobbled up by Gold Coast developers. By 1962 she and several Brisbane friends had decided that the state government's drive for development demanded a more proactive campaign, one which could educate the Queensland public in conservationist principles. In September of that year, she and three associates formed the Wildlife Preservation Society of Queensland, with a mission, through its journal *Wildlife*, to promote "the preservation of all forms of . . . flora and fauna and the education of children and adults in the principles of wildlife protection and conservation."[15]

Judith had little time to think about the Great Barrier Reef during the tenuous early years of the society's existence, when the journal was short of money and without a professional editor, yet the Reef's unique beauty had lodged more deeply in her consciousness than later opponents would claim. In 1949 she and Jack had stayed for several weeks at Lady Elliot Island on the southern boundary of the Reef, sharing the lighthouse keeper's cottage with Jack's daughter. Shocked at the evidence of destruction on the island caused by goats and guano miners, Judith was nevertheless enchanted by its fringing coral reef: "I wandered over it amazed at the colours of the coral, the shellfish and the tiny darting fish and crimson and blue slugs and stars and clams in its pool-gardens, and stared down from a small boat at its shelfs and coral crabs. I fell in love with the Reef then . . ."[16]

The sublime architectural creations of these tiny coral polyps became her personal symbol for how brave individuals like Jack McKinney worked, for they were like coral "Builders," creating structures of moral value and defiance in the face of oceanic forces of destruction:

Only those coral insects live
that work and endure under
the breakers' cold continual thunder.
They are the quick of the reef
that rots and crumbles in calmer water
Only those men survive
who dare to hold their love against the world;
who dare to live and doubt what they are told.
They are the quick of life;
their faith is insolence; joyful is their grief.[17]

It was soon after setting up the Wildlife Preservation Society that Judith met Len Webb. The rain forest ecologist was another lone "Builder" and poetry lover, and was keen to join her cause. As she got to know him, Len proved to be "a vital and urgent man with a love for the magnificent forests he studied, [who] travelled to and fro, talking to people and making himself unpopular, but also being heard by those with foresight."[18]

Lean and leathery, with a thin mustache, a tanned complexion, and spectacles like navigator's goggles, Len was five years older than Judith and, like her, the product of a rural family, although by upbringing he was closer to the shearing shed than the homestead. Born to a station-cook mother and a horse-breaker father, he'd grown up acutely aware of "the gulf between . . . the squatters and the station hands." But there were no gulfs between Builders, and Judith delightedly persuaded him to become the society's vice president.[19]

Len's scientific education had been hard won. He began his working life as a junior clerk in the Queensland Department of Agriculture, before becoming a trainee chemist at the state herbarium. After obtaining his high-school certificate at night school, he enrolled as an evening student at the University of Queensland, where his politics led him to the Radical Club, and his chemistry to wartime appointments in munitions factories in Melbourne and Sydney. The army also gave him the opportunity to study biochemistry at night at Sydney Technical College. Then came a life-changing job offer from the Commonwealth government. In a phytochemical survey of Australian rain forest plants prompted by a shortage of drugs, Len was to look for specimens containing chemical alkaloids suitable for medical use.[20]

Over the years 1944–52 Len developed a passion for the towering rain forests that skirted the Barrier Reef. With thousands of unknown plants to study, he learned to test for alkaloids using every tool available, including his formal knowledge of botany and biochemistry, a knack of tasting alkaloids on his tongue, and an arsenal of folk wisdom acquired from Aborigines, timber workers, and tin miners. Around campfires he discovered that many of these tough, solitary men shared his sense of "the mystery and sacred beauty of the rainforests."

One day while working in the forest, he experienced a type of epiphany that produced an uncanny feeling of intimacy, or "cathexis," with the surrounding flora and fauna. He was suddenly "overwhelmed, without feeling claustrophobic or afraid, by this complex terrestrial ecosystem." In later conversations with Judith Wright and John Busst, he likened this enchantment to being infused with "soul," "heart," or "spirit."[21]

As Len and his self-taught assistant, Geoff Tracey, traveled through the Reef forest, they had to devise ways to categorize the vegetation types they encountered, and to understand their patterns of growth and association. Groping attempts to find a suitable language led them both to the nascent science of ecology. Len undertook a pioneering doctoral degree in the subject at the University of Queensland, and then persuaded the Commonwealth Scientific and Industrial Research Organisation (CSIRO), his employer, to fund a two-person rain forest ecology unit based in Brisbane. Len and Geoff's daunting first task was to map the diversity of Australia's rain forests from Cape York to Cape Otway, in Victoria, and to discover what caused the varying patterns of community and distribution.

It was an awesome responsibility, and one that engaged all of Len's emotional and scientific sensibilities. As he later liked to say: "Trees have no blood banks to succour them after fire and mutilation, yet without the green stuff of their sap, there would be no redness in animal blood, no sun's energy and no life for us who cannot dine on dust. A tree is a magic creature, whose ancestors are lost in the mists of time."

He and Geoff identified so deeply with the rain forests of the Reef that they came to see themselves as custodians of a natural heritage that seemed everywhere under siege. Heretically, they'd arrived at the now orthodox view that Australian rain forests were unique fragments of the ancient vanished continent of Gondwanaland, an idea that generated heated arguments with development-minded CSIRO colleagues who wanted to clear the northern forests for agricultural and pastoral use. As a result, Len claimed, he woke up one morning in the early 1960s to find himself "an adversary of the powers that be." He'd become a radical conservationist.[22]

Like most converts, Len wanted to evangelize. One of his earliest targets was John Busst, whom he'd met in the late 1940s when the artist sent him a poisonous shrub from Bedarra Island for testing. The two hit it off at once. On Len's regular visits to the island, they discovered a common "avidity for scientific-romantic ideas," as well as a passion for Tennyson's poetry, Wagner's music, and the euphoric effects of ethanol.[23]

By birth and education, the two men were poles apart. A decade older, Busst was born in 1909 in Bendigo, Victoria, into the family of a wealthy mining warden. He went to elite Wesley College, then to the University of Melbourne, and was intended for a glossy legal career like his friend Harold Holt. Instead, something impelled him in the early 1930s to drop out of university and become an artist. Aligning himself initially with the womanizing painter Colin Colahan, he switched allegiance to the ascetic arts-and-crafts guru Justus Jorgensen, whose Montsalvat artists' colony in Eltham, Victoria, was modeled on the communal ideals of William Morris and Leo Tolstoy. Busst had limited talents as a painter, but he was able to use subsidies from his father to sustain a bohemian lifestyle throughout the Depression, and to learn the skills of working with adobe, wood, glass, and stone.[24]

A bequest from his father allowed Busst to follow in the footsteps of the talented painter Noel Wood, who lived on Bedarra Island. Enchanted by a short holiday on the island, John and his new wife, Alison, eventually bought half of Bedarra, where they stayed for seventeen years, building a mudbrick house, laying out a tropical garden, selling the odd painting, and living a hedonistic life. Like the Banfields, though, they were alienated by the growing intrusion of tourists. In the mid-1950s John published his first conservationist article—as it happened, on Banfield's favorite subject, Bedarra's breeding caves of gray-backed swiftlets.

Two years later the couple decided to move to Bingil Bay, where they hoped to buy enough land to insulate themselves from developers and "Philistines." John told Patricia Clare that "the days for living on tropical islands have gone. Half Bedarra is now a resort and there's no privacy any more. Speedboats everywhere, buzzing all around you."[25]

Yet nowhere in the Reef province seemed sacrosanct. Even their new back garden at Bingil turned out to be at risk. A patch of rain forest at Clump Point, just behind their bungalow, was suddenly co-opted by the army for exercises in jungle warfare and defoliant bombing. Len Webb happened to be staying on the night the Bussts decided to fight this army takeover. He agreed that there were sound ecological reasons for protecting the forest, but suspected that the Bussts didn't yet know them. They just "wanted to keep it virgin." Len offered to instruct them in ecological principles, an education that ultimately helped the couple to rescue rain forest from the army at McNamee Creek as well as at Clump Point.[26]

These small victories fired John's passion for conservation, to the point where he began to press for protection of the forest, river, and reef habitats of the whole Innisfail–Tully region, an area under severe pressure from prospective cane farmers. Gradually John's small branch of the WPSQ widened its remit to campaign against chemical weed and tree killers, and in favor of government rebates for landholders who spared natural vegetation along watercourses. In conversations with Patricia Clare he was adamant that the reefs and rain forests were critically interdependent, because "the clearance of vegetation affected the coast's relationship to the Reef waters alongside it. The river estuaries, mangrove swamps and shallow waters close to the coast supported life that was part of the whole marine system. These waters must be affected when vegetation was cleared and torrential rain stripped soil from the coast and dumped it into the sea."[27]

As the friendship between Judith Wright, Len Webb, and John Busst deepened, they increasingly exchanged ideas and tactics. Even so, becoming "welded in a very deep companionship" wasn't always easy. By inclination as well as by upbringing, John Busst was less democratic and tolerant than the other two. His "Irish blood" and the urgency of the cause often made him impatient. He liked to flaunt the belligerent nickname of "The Bingil Bay Bastard," scorned the idea of "controlled exploitation," and believed it was necessary to force environmental change on ignorant "Philistines." All this made him chafe at his two friends' commitment to slower, educative methods and long-term democratic goals.[28]

In November 1966, for example, he ranted at Len Webb about "the

puss-footing, polite, ivory towered vapourings" of Webb's academic associates in the ACF. "Look, chum, just forget about the education angle for
a while, will you? It's just bloody silly for the moment to educate future
generations to preserve the rain forest—there won't be any." Judith went
some way toward shifting his prejudice, however, when she linked John
up with Barry Wain, a sympathetic young journalist on *The Australian*
who agreed to publicize his prolific flow of letters.[29]

In 1966 Len Webb presented a talk to a group of student teachers
which, when it was subsequently published by his two colleagues in their
journal, *Wildlife*, also helped them to understand ecology—"the science
of relations between living things and the landscape." John and Judith
learned from Len that reefs and rain forests were interdependent communities, or ecosystems, which couldn't be exploited piecemeal without
damaging the whole. Len advanced a scientific case for something Judith had always believed—that Aboriginal peoples had achieved a "precarious but stable equilibrium" between conservation and resource use,
which was "blown apart" by the European introduction of sheep and
cattle and the wasteful practices of overgrazing and tree clearing. These
drastic environmental interventions, Len said, had triggered unplanned
and unstoppable chains of change.

Len quoted lines from one of Judith's poems to preface an argument
that wilderness and nature reserves answered deep emotional and aesthetic yearnings within human beings: "we need," he reflected, "a new
word to convey that feeling of deep enjoyment and wonder, that feeling
of privilege in witnessing the life of wild animals and birds, and in moving among wild scenery." Yet he cautioned against being "sentimental"
about nature, a point he made regularly, until Judith advised him sharply
one day "that I should not be ashamed of 'sentiment . . . It's a good
word—look it up in the dictionary.'" The definition, "thought tinged
with emotional feeling bound up with some subject or ideal," surprised
and delighted him. Nothing could be closer to his own views. Her intervention, he later claimed, had introduced him to the crucial concept of
"emotional intelligence."[30]

An article Judith wrote soon after this, "Conservation as a Concept,"
revealed the reciprocal impact of Len's ideas. She attacked science for
having "separated man from nature" by confusing objective methods of
investigation with an abandonment of "any true and deep consideration
of moral ends." However, she continued, "a hopeful new science has

arisen in the new studies of ecology, which are moving into the human as well as the biological fields." She believed that a fusion of conservationist ideals with ecological methods could even fulfill the poet-scientist Goethe's centuries-old plea "for a science of living experience": "[It] does represent a groping movement towards a new kind of understanding which shall take into account actual living processes and interdependencies, and can see man as part of a wider process and subordinate to its laws."

Such a fusion might also bridge one of Western civilization's most tragic divides, providing "a point at which a new spark can perhaps jump across the gap that at present separates the arts and the sciences—to the great detriment of each—and allow a new kind of cooperation and understanding to grow up between men."[31]

Though emboldened by their collaboration, none of the trio underestimated the enormity of the task ahead. The Queensland government had openly signaled its intention of allowing petroleum companies to explore for oil and gas anywhere within an area encompassing 80 percent (eighty thousand square miles) of the Great Barrier Reef. Stopping them looked almost impossible: it was a "David versus Goliath" confrontation. As a "fringe" group of unpaid and powerless amateur volunteers without government recognition, Judith recalled, "we were opposing wealthy interests, entrenched government policies and political forces that seemed immovable."[32]

In Joh Bjelke-Petersen and his government, they were pitted against one of the most ruthless and effective populist governments in modern Australian history. The premier was adept at exploiting Queenslanders' suspicions of southerners and "interfering" federal governments, and he grabbed every opportunity to represent conservationists as "a lunatic fringe" of "nitwits," "cranks," and "rat-bags" (eccentrics). Untroubled by consistency, he and some of his ministers also accused the campaigners of being flower-sniffing sentimentalists, "commies" intent on overthrowing the Australian way of life, and agents of American capitalism aspiring to plunder their state.[33]

Bjelke-Petersen was equally quick to exploit any chink of division within the conservationist cause, relishing the naïveté of the leaders of the ACF and the GBRC, who believed that appeasing the state government and the oil companies would produce the "responsible" partitioning of the Reef for mining. Afraid of the electoral effects of Queensland

National Party populism, successive Liberal federal governments had been too nervous to challenge the state's shaky constitutional claims to own the Reef. In twin parliamentary acts of 1967 the federal and Queensland governments agreed to "co-operate for the purpose of ensuring the legal effectiveness of authorities to explore for, or to exploit the petroleum resources of those submerged lands."[34]

Lack of detailed biological research into the ecology of the Reef played further into Bjelke-Petersen's hands. Without such expert evidence, the WPSQ conservationists could do little but call for a ten-year moratorium on mining, and urge funding for a federal marine research station in the region. Neither suggestion troubled Bjelke-Petersen, who shrewdly exploited long-standing disciplinary divisions between Australian geologists and biologists on the issue of mining the Reef.

Never one to miss a trick, Bjelke-Petersen appointed an American geologist, Dr. Harry Ladd, to undertake a survey of the potential impact of mining on the Reef. Ladd, a man with extensive mining experience, managed to achieve this mammoth task in less than a month—flying over much of the area in the company of officials from the Queensland Mining Department. As the state government had hoped and the conservationists feared, Ladd in his report of March 1968 considered the outlook for oil and gas discoveries to be "promising." He further recommended that "non-living" parts of the reef should be developed as sources for agricultural fertilizer and cement manufacture.[35]

A furious Judith Wright likened this to using the Taj Mahal for road gravel, and John Busst urged his journalist friend Barry Wain to spread the word that Ladd's report was "scientific nonsense." At a major ACF symposium of scientists, government officials, and oil miners in 1969, Australian Museum director Frank Talbot observed that geologists tended to believe no harm could come to the Reef from any activity, while biologists "were less confident because they were aware that living matter was more fragile and sensitive than geological matter." Several of the geologists' papers scoffed at "hysterical" conservationist claims that oil spills might damage corals and other marine organisms. Well-publicized quotations from a second American geologist that the Reef should be "exploited immediately, and to the hilt" also delighted the Queensland government. Given that "oil companies and the like" were funding most of the Reef geologists' research, Judith Wright found such views unsurprising. John Busst simply wrote off the whole symposium as "a bloody shambles."[36]

Academic disdain for the fledging discipline of ecology was a further obstacle. Such attitudes weren't unique to Australia: the Canadian ecologist David Suzuki has admitted that when he graduated as a geneticist in 1961 he regarded ecologists as the kind of people who strolled down the road listening to birds and calling themselves scientists. Geoff Tracey and Len Webb experienced similar prejudice from CSIRO colleagues, on the grounds that they were field-workers who used "unrespectable" methods. "Instead of a null hypothesis," Len explained, the community-based ecology (synecology) that he practiced, "requires the generation of a new hypothesis—which might or might not be amenable to later test—in a mental procedure more closely akin to the balanced personal judgment of an English history scholar." Like historians, the ecologists depended on "the inductive synthesis of fragments of evidence from many sources" for their interpretations.[37]

This suspicion of ecology as "the subversive science" only began to lessen, Len thought, when he could use computing tools to create simulation programs able "to synthesize complex systems from a multitude of interacting parts."[38]

During the torrid conservationist campaigns of 1968–69, such respectability was still a long way off, and influential supporters hard to find. The odds were worsened by a succession of personal setbacks. Judith, still suffering from Jack's death in 1966, was also troubled by advanced hearing loss due to a childhood disease. Her close friends Arthur Fenton and Kathleen McArthur both suffered collapses trying to cope with the heavy workload of WPSQ business. John Busst, devastated by the disappearance and probable drowning in December 1967 of Harold Holt, confided to friends that he had throat problems, which were feared by his doctor to be the onset of cancer.

Fortunately, sickness didn't diminish John's energy. After Holt's death, John wrote to American President Lyndon Johnson to solicit support for a series of "Harold Holt Commemorative Marine Reserves," which would culminate in a Great Barrier Reef Marine Park. Starting at the top as usual, he also lobbied the new federal Liberal prime minister,

John Gorton, and the leader of the opposition, Gough Whitlam. Having cornered both of them while they were vacationing on the Reef, he followed up with a barrage of letters.

Citing a newfound academic ally in Cyril Burdon-Jones, professor of zoology at Townsville University College, Busst stressed that the need to save the Reef transcended adversary politics: "it was a matter of international concern, too important to be made a political football or subject to parochial state interests." He urged Gorton and Whitlam to test in the High Court a recent opinion of constitutional legal expert Sir Percy Spender that Queensland had neither domestic nor international sovereignty over Reef waters.[39]

Even with the heroic support of John and Len, Judith Wright was sometimes brought to the edge of despair under the cumulative weight of so many handicaps, her own and the cause's. This was reflected in her poem "Australia 1970."

For we are conquerors and self-poisoners
more than scorpion or snake
and dying of the venoms that we make
even while you die of us.[40]

"If the Barrier Reef could think it would fear us," she fretted, ". . . we have its fate in our hands." Every day it experienced a bombardment of human-generated toxins. Following her reading of Rachel Carson's *Silent Spring*, she believed that oil and gas would simply be the coup de grâce, given the damage already done by fertilizers, pesticides, dredging, sugar-plant effluent, and urban sewage.[41]

Somehow, she and her friends had to persuade Australians to feel that the Reef belonged to them, and they to it—in the same way that coastal Aborigines had thought about and cared for the Reef in centuries past. European Australians needed to see it as a core symbol of identity, or at least as the fabric of a new and affirmative Australian myth. The Reef had to be something that plumbed the deepest reservoirs of Australian imagination, intuition, and knowledge; it had to be seen as more profound and urgent than any temporary accession of material wealth. Conservationists had to show their countrymen that the struggle for the Reef went far beyond stopping greedy state politicians and reckless resource companies: it was a clash between good and evil, between "life and death."[42]

To Judith, the Reef's fate was "a microcosm of the fate of the planet. The battle to save it is itself a microcosm of the new battle within ourselves." The Reef is "a symbol," she later wrote, "of humanity's failure to recognize its responsibility and of the whole relentless process of commercialization and industrialization, pollution, self-interest and political importance." Len agreed that the Reef was a bastion whose fall would unleash orgies of deforestation and oceanic plundering all around the world. John Busst foresaw the Reef becoming "a quarry surrounded by an oil slick."[43]

Australian conservationists were fighting for nothing less than the biosphere—"the thin covering of living organisms supported by earth, air, water and sunlight." To win such a war would be epochal; it would be an expression and a marker of Australia's having at last become "a country of the mind" rather than a haunt of predators and a shrine of Mammon. A country where the Aboriginal ethic of biocentrism and stewardship of nature had at last taken root, and where, Judith wrote, a "re-imagining of nature" could encompass "the arts, affirming the truths of feeling, and the sciences, affirming the truths of intellect."[44]

Though Judith and her friends did not yet know it, the tide of war had already begun to turn. In part this was due to pure chance. The 1967 *Torrey Canyon* oil tanker disaster in Britain had generated ripples of negative publicity for the global oil industry, and this turned into a torrent in late January 1969 when a rig started spewing crude oil into the blue waters off Santa Barbara, on California's coast. Nightly television programs showed infernos of rotting fish, oil-clogged seabirds, and viscous black beaches. Bjelke-Petersen's bleat of "Don't you worry about it; it won't happen in Queensland" sounded increasingly hollow in the face of American citizens publicly pointing out that the same assurances had been made in Santa Barbara. Looking back, Judith Wright believed that the Santa Barbara disaster was the watershed moment when popular sympathy in Australia began to flow the conservationists' way.[45]

While the Reef campaign was starting to attract new support, its enemies were for the first time beginning to take some direct hits. One of the most telling was a newspaper revelation that Bjelke-Petersen and some of his ministers had invested heavily in the oil companies they'd licensed. Some risible public utterances by the Queensland minister for mines didn't help their case either: Camm gave a speech arguing that oil, being protein, would actually provide nutritious food for fish and other marine organisms.

Toward the end of 1969 a series of local polls conducted by the WPSQ at rural shows and festivals and in urban shopping malls showed that a majority of Queenslanders now opposed the idea of mining the Reef. Around the same time, a bipartisan group of politicians and citizens formed the Save the Reef Committee, chaired by outspoken federal Labor parliamentarian Senator George Georges. Despite his growing unease, Bjelke-Petersen insisted that oil exploration must go ahead because the licenses couldn't be rescinded without serious financial loss.

The conservationists' cause took a decisive turn at the beginning of the new decade when John Busst's lobbying of trade unions finally produced results. In early January 1970 George Georges warned the consortium of Japex and Ampol petroleum companies that their survey ship the *Navigator* would be placed under a union black ban when it reached Repulse Bay. Powerful national and local unions also declared their support, both for making the Barrier Reef a marine reserve and for mounting a legal challenge to Queensland's claim to Reef sovereignty. On January 13 Ampol announced the suspension of its Repulse Bay survey, and a month later their Japanese affiliate, Japex, canceled the contract. "Now at last, the breakthrough has come," John Busst declared. "It has taken two and a half years to find the weapon."[46]

Within a month the federal government agreed to dedicate three million Australian dollars for an institute of marine science at Townsville, to study "ways of researching and protecting the Great Barrier Reef"—the need for which was underscored when news broke soon after that a fully laden Ampol oil tanker had been holed near Tuesday Island in the Torres Strait. Out of the ship's hull oozed a six-mile oil slick that killed hundreds of birds and fish and impaired the livelihood of Islanders for years to come. Backed into a corner by popular outrage, Bjelke-Petersen was forced to agree to an inquiry into the mining of the Great Barrier Reef—later upgraded to a royal commission. On April 16, 1970, Prime Minister Gorton introduced a bill into federal parliament claiming Commonwealth sovereignty over the Reef and its waters.[47]

At this point the Great Reef War was effectively won. The election of the Whitlam Labor government in 1972, the affirmative report of the royal commission, and the High Court's decision in favor of Commonwealth sovereignty all paved the way for an eventual settlement. Both the cautious ACF and the pragmatic GBRC now strongly supported the formation of a marine park. Patricia Mather, secretary of the GBRC, even consulted with Judith and her WPSQ "stirrers" when drafting plans for its governance. Judith and her colleagues were unenthused by Mather's concept of a "multi-use" marine park, which was nevertheless incorporated in the Whitlam government's Great Barrier Reef Marine Park Act, passed in 1975 but not fully promulgated until 1979. The act prescribed that most of the Reef province become a marine park governed by a committee of the Great Barrier Reef Marine Park Authority, which was in turn answerable to a federal government minister.

Finally, on October 26, 1981, the Great Barrier Reef received what two of its finest historians, James and Margarita Bowen, have called "a conservation climax"—World Heritage listing "as the most impressive marine area in the world." The Reef met all four of UNESCO's "natural criteria." It was an outstanding example of the earth's evolutionary history, an arena of significant ongoing geological processes and biological evolution, a superlative natural phenomenon, and a significant natural habitat containing threatened species of animals or plants with exceptional universal scientific value.[48]

John Busst had prepared materials to take before the royal commission, but he didn't live to present them. Early in 1971, at the age of sixty-one, he quietly "dropped off his twig," as Len put it. Judith composed a simple tribute for his Bingil Bay memorial: "John Busst / Artist and lover of beauty / Who fought that man and nature might survive." His two friends-in-arms missed him sorely.[49]

Judith died in 2000 as one of Australia's finest twentieth-century poets and was widely mourned. Len left his beloved rain forests behind eight years later, having been much awarded for services to their ecology. Both had remained proud of the success of their small band of amateurs in helping to win the Reef war, even though they realized that the victory would always be provisional.

Judith ended the twentieth-anniversary edition of her *Coral Battle-*

field (1996) with a chapter called "Finale Without an Ending": she feared that the looting and poisoning of the Reef was continuing. Popular memory was short, and onslaughts on the Reef's oil and mineral resources could resume anytime. She was deeply troubled, too, that Aboriginal people had emerged from the war with nothing to show for the loss of their heart's country.[50]

The trio died without having achieved that vital fusion between the emotional arts and the intellectual sciences they'd wanted. It had taken fire in their minds, but not in the country as a whole. Judith was even publicly lambasted in 1971 by a professor of agricultural engineering for using "lay conservationists" and "emotional arguments" to defend the Reef. "I will go in and try to arouse public feeling, and have done, and on this my conscience is more than clear," she replied bitterly.[51]

The still-yawning gulf between the arts and sciences would one day cost the Reef dearly, and that time was not far off.

12

EXTINCTION

Charlie Veron, Darwin of the Coral

SIR DAVID ATTENBOROUGH stands at the lectern of the Royal Society in Carlton House, London, on July 6, 2009, about to introduce the afternoon's speaker. Printed below him are the words "Celebrating Three Hundred and Fifty Years," to remind us that this august "Society for Improving Natural Knowledge" was founded around 1660 and is among the oldest learned scientific bodies in the world.

The video of the occasion catches a ripple of expectation among the audience. It suits the lecture's confrontational title: "Is the Great Barrier Reef on Death Row?" Then, in that familiar tone of breathy intimacy, Sir David introduces J. E. N. Veron, former chief scientist of the Australian Institute of Marine Science. "But," says Sir David, smiling broadly, "I'll call him Charlie, a name he carries because he shared Mr. Darwin's obsession with the natural world." Without specifically saying so, Sir David is telling us that we are about to hear from a modern-day Charles Darwin.[1]

Many of the scientists in the room already know how apt this comparison is: there are uncanny resemblances and intellectual links between today's speaker and the Royal Society's greatest-ever fellow, Charles Darwin. All Charlie Veron's friends know, too, that against the countervailing conditions of the modern world, he has made himself an

Charlie Veron (Courtesy of Charlie Veron)

internationally famous scientist without ever losing Darwin's fierce independence, unquenchable curiosity, and passionate love of nature.

Charlie, Sir David continues, is one of the world's greatest scientific authorities on corals and coral reefs. He's discovered and described more than a third of the known coral species and produced definitive catalogs of all the world's corals. For much of his life Charlie has been blessed with the most enviable job imaginable. He's swum among the reefs of the world to explore coral cathedrals of "beauty, wonder, and astonishment" that most of us know only from our television screens.

But today—Sir David's voice takes on a somber note—Charlie comes with a different task: to show us how coral reefs are the keys that can unlock the truth about the bewildering changes we've unleashed in our climate. Perhaps he may answer the question that nags at us all: Do the reefs tell us that the future is worse than we realize?

When the applause subsides Charlie walks to the lectern, a wiry, tanned figure wearing a red shirt and dark jacket. In his husky Australian voice he thanks Sir David, and then answers his concluding question. "Yes," he says bluntly, things will be worse than most of us realize, and so, sadly, "I will not be telling a happy story." For the next forty minutes he tells the silent audience why the Great Barrier Reef and all the world's reefs face a likely mass extinction within the life span of the younger listeners present.

At one level this Royal Society lecture—and the book that underpins it, *A Reef in Time: The Great Barrier Reef from Beginning to End* (2008)—marks a shift in theme and tone for a man who has written so joyfully about coral reefs. For forty years Charlie has celebrated their astonishing multiplicity and complexity. Now we watch him focusing all his intellect and passion to prophesy a reef apocalypse. It's obvious how much he'd like to avert what he predicts. To have any chance of this, though, Charlie must answer the skeptic's question: How do you know? And then its brutal follow-up: Why should we care anyway?[2]

Here is how Charlie Veron came to know about the prospective extinction of the Reef, and why I think he thinks we should care.

One day in 1951, John Veron, a second-grade boy of six at East Lindfield primary school, in the bushy northern suburbs of Sydney, presented his nature-loving teacher Mrs. Collins with his contribution to the ritual of show-and-tell. Holding up to the class his tableau of dead frogs, spiders, and wasps fastened onto cardboard, she admiringly dubbed him "Mr. Darwin." A few weeks later, when ordering him to take a jar of putrid marine worms out of the classroom, she reinforced the title, "Mr. Charles Darwin." His classmates, with the usual flair of children for seizing on difference, instantly turned this into the nickname Charlie, chanted with a rising pitch on the first syllable and a falling inflection on the second—Char*lie*, Char*lie*. Though neither he nor his taunters had any idea who Charles Darwin was, the boy was branded for life.

Mrs. Collins, to her credit, didn't lose interest in her protégé. Hearing later of Charlie's concern that a newly published text, *Australian Seashores*, had failed to mention some of the marine worms in his collection, she urged the boy's mother to take him to visit the amiable coauthor of this book, Isobel Bennett, at the University of Sydney's zoology department. Awed by the tall, white-coated lady wearing huge spectacles and smelling of formalin, Charlie was ecstatic when she used his nickname and praised his knowledge of the specimens tipped into her enamel tray.[3]

We may guess that Isobel Bennett felt some instinctive affinity with this earnest, self-taught little boy. She loved the seaside world, and was a woman in a male-dominated department who'd begun her career as a research assistant with no formal qualifications, only her amateur enthusiasm. Their meeting was portentous. "Issie" would become as loyal a friend and patron to Charlie as the kindly Cambridge botanist John Henslow had been to Charles Darwin.[4]

But these were Charlie's last positive memories of school. Like Darwin, he was moved for his middle and senior years to a more famous private establishment that made his life a misery. Charlie's experiences at Barker College in Sydney verged on the traumatic. He slumped to near the bottom of the class in most subjects, loathed what he perceived as the pompous senility of the teachers, and acquired asthma and a stammer. His retreat into himself was interrupted only by flares of temper at bullying schoolmates and by covert rebellions against his teachers.

The boy didn't even compensate with especial prowess on the sports field: Charlie was indifferent to the manly activities of football and cricket. His ex-brigadier father had tolerated his son's love of nature, but he couldn't hide his disappointment at all these signs of the boy's inability— an attitude sometimes echoed by Charlie's otherwise loving mother. Even now, Charlie cannot recall a single word of praise from either parent throughout his childhood.[5]

One antidote to the misery didn't last. Young Charlie developed an intense reverence for the designer God who'd given mankind the lovely gifts of nature. At the age of thirteen, however, he experienced a counter-conversion. A classmate lent him a glossy journal article on the missing link, which summarized Darwin's hypothesis that humans had evolved naturally, without divine assistance, from apelike ancestors. Charlie's rational mind, steeped in naturalist knowledge, recognized the idea as irresistibly true. He was appalled that it had been kept from him all these years. Tearful confrontations with the school chaplain and headmaster confirmed the conspiracy. After being confined to home for several weeks to regain his mental health, he returned to school as an implacable evolutionist.[6]

Like Darwin, Charlie drew his deepest consolation from an inborn love of nature. What others saw as a hobby, he knew to be a lifeline. Nature offered him everything: sensual and aesthetic delights, physical tests and thrills, intellectual and imaginative mysteries, and emotional escape and comfort. He found peace and pleasure walking the beaches of Collaroy, searching the rock pools of Long Reef, riding his bike along the bush tracks of Ku-ring-gai Chase reserve, wrestling his dog Jinka, labeling specimens for his garage museum, and hand-feeding bits of crab to his blue-ringed octopus, Ocky, in happy ignorance of the creature's lethal venom.[7]

Just as Dr. Darwin had decided that his son's pedestrian academic record made him suitable only to become a Church of England clergyman, so Charlie Veron's parents decided that the boy's poor school results destined him to become a bank clerk. Chance—that crucial element in natural selection—rescued both boys. Charles Darwin's opportunity came when his friend Henslow was asked to find a gentleman companion to entertain naval captain Robert Fitzroy on the HMS *Beagle*'s circumnavigation of the world. Charlie Veron's luck was more outlandish still. The New South Wales government had decided on this single occasion, in the final year of Charlie's schooling, to indulge an experiment by an educational psychologist to test whether IQ results could predict future university success better than did school-leaving exams. To Charlie's astonishment he topped a select group who were subjected to a rigorous series of competitive tests. This earned him a scholarship to attend any university in the country.[8]

Charlie Veron would have loved Darwin's chance of an informal scientific apprenticeship sailing around the world, exploring wild lands and investigating puzzles of nature. Given that those days had passed, he opted for a modern equivalent. To the bafflement of his scholarship administrators, he asked to enroll as a science student at the small rustic University of New England in Armidale, New South Wales. The university was situated in the middle of the crisp, sunny country that had imprinted itself on Judith Wright's imagination, and it helped to heal Charlie. His asthma vanished and he found he could run for miles through the bush with exhilarating freedom. He made boisterous friendships at his all-male college, he discovered the sublime music of Schubert and Liszt, and he joined the university's Exploration Society.

> I loved being alone in the wilderness, going long distances, constantly talking to myself in my mind, thinking about all I saw, and soaking up

the beauty of one place after another. Every day I would make a point of finding a quiet spot, preferably beside a stream, sitting down, remain absolutely still, and gradually let myself stop thinking. I would be aware of the light on the water, maybe a dragonfly hovering or a cricket chirping, but I wouldn't think about it. This might last for twenty minutes, then a thought would come into my head, and I would come to as if out of a trance. I would get up relaxed and rested as though I had slept for hours.[9]

An ardent sea lover, Charlie overcame the disadvantages of living inland by hitchhiking to the nearest islands at the southern end of the Great Barrier Reef. On his first trip to the Reef he stayed at beautiful Heron Island, where he was proud to show some French underwater photographers where to find giant manta rays on the sandy floor of the Wistaria Reef channel. A few years later he followed up this initiation by camping with ten student friends for two weeks in the solitude of Zoe Bay, on the forested island of Hinchinbrook, near the Palm Island group. They swam naked in the warm clear waters and tested their courage by climbing the steep side of the island's soaring 3,675-foot-high mountain.[10]

Apart from a newly arrived associate professor, Harold Heatwole, the university's department of zoology offered little direction or guidance. This suited Charlie perfectly: he treated formal research requirements as chances to extend his recreational naturalism. Studying glider possums for his honors thesis introduced him to the eucalyptus forests of Dorrigo and to the bush skills of its foresters. Working on the temperature regulation of lizards for a master's degree was less stimulating, but he earned some part-time income collecting snakes for the department—at least until he secretly released all the snakes, rather than see them die from the defective conditions of their confinement.

Unusually, he was also hired as a teaching fellow to run the diverse courses of academics who were away on sabbatical. Again, Charlie didn't mind: teaching across the whole departmental curriculum, including genetics and physiology, proved an unexpected joy. Even so, he felt devoid of scientific ambition—apart from a hazy longing to imitate Darwin's adventurous colleague Alfred Russel Wallace by exploring the Amazon.[11]

Falling in love with a fellow student, Kirsty Mackenzie, deferred Charlie's jungle dream. Wanting to remain at university with her, he began to scratch around for a PhD topic that might attract three further years of university funding. While jogging one misty morning he noticed that blue dragonflies had turned gray in the dawn. As the fog lifted, he watched them "orient themselves to the rising sun" and turn back to blue. Charlie had found his subject. The university's head of entomology was known to love dragonflies, so Charlie grabbed his chance. He became an entomologist, with enough funding and equipment to develop a thesis. He taught himself to monitor dragonfly eye functions with an electron microscope; to record changes in color, temperature, and behavior; and to use a freeze microtome to examine minute slices of frozen tissue. Balancing this intensive lab work with time in the field, he spent hours at a lagoon just outside of Armidale, observing his swift-flying subjects.[12]

But dragonfly work, though intriguing, didn't satisfy Charlie's craving for exploring nature. Early in 1967 he decided to escape thesis tedium by taking a rudimentary scuba diving lesson. With the shouted advice of "Don't come up too quickly or hold your breath" ringing in his ears, he plunged into the sea off Cronulla in southern Sydney, to discover a new state of being: "The dive was fantastic! I could see at least six foot, mostly rock and mud, but the feeling of being weightless in a hidden world was thrilling and a little frightening."[13]

Other life-changing experiences quickly followed this underwater epiphany: marriage to Kirsty in 1968 was followed by the joyful birth of a daughter, Noni, in 1970. Yet Charlie's hypnotic new hobby stuck through all these excitements, to the extent that he and Kirsty spent part of their honeymoon gazing at corals off Heron Island.

Two years later Charlie took a formal diving ticket—a qualification required in order to scuba dive—and then set up a scuba club among his fellow students. The half-dozen founder members based themselves at Coffs Harbor, where the university's zoology department owned a fibro shed and a fourteen-foot tin runabout. Cramming themselves into the boat one day, they decided to expand their horizons by taking the one-and-a-quarter-mile journey to the Solitary Islands. Here, diving off South West Solitary, Charlie and his friends made their first marine discovery. Nestled beneath the rocky shore, in a zone assumed to be well south of the tropics, they found an embayment of tropical corals "growing in abundance." Perhaps influenced by the feats of Eddie Hegerl's student divers at Ellison Reef, they decided to survey the Solitary Islands corals as a way of pressing for the area to become a marine nature reserve.[14]

The rudimentary nature of their diving techniques and their slender knowledge of marine species marred the mission, yet it would have incalculable implications. The student who'd been allocated the task of collecting the corals had to drop out, so Charlie took his place. Without realizing it Charlie had found his future subject. Not long after this, at a time when his dragonfly thesis seemed in the doldrums, he rather desperately cited this amateur coral survey in response to an advertisement from James Cook University (JCU), Townsville, for a postdoctoral scientist with scuba diving abilities to work on the corals of the Great Barrier Reef.[15]

While he was waiting to hear back from James Cook University, the disparate strands of his PhD thesis suddenly came together, and in 1972 he found himself at the Fourteenth International Congress of Entomology in Canberra receiving a prize for the best student paper, based on the findings of his dragonfly work. Four offers of postdoctoral scholarships immediately followed, including, to his astonishment, one from James Cook University in answer to his application: Charlie had been the only candidate. A prestigious Canadian university led the field of scholarship offers with an enticing package, so Charlie accepted, writing half-dazed letters to decline the others. Suddenly he'd metamorphosed into respectable Dr. J. E. N. Veron, entomologist and academic-to-be.

Then the doubts began to surface. Did he really want to join the kind of university environment where postgraduates had to compete fiercely against one another, as well as guard against their ideas being stolen? And the thought of losing his independence was horrible: he'd "just be a cog in a machine and not free to follow my own nose." Thinking guiltily that he should at least take JCU's advice to complete his Solitary Islands coral report, Charlie visited the University of New England library to find only one pertinent book on the shelves. It was *The Great Barrier Reef*, written by his old mentor Isobel Bennett. As he flicked through the pages, he couldn't help thinking that "it was full of photos of the place I had just said I was not going to work in, work that would be all mine, in a place that was so important, and so beautiful and so challenging."

He quickly scribbled two letters, to Canada and to Townsville, each asking to reverse his previous decision. With this one impulsive action, Charlie Veron triggered three bizarre outcomes. He was assailed by "a blubbering barrel of fury" for giving up a career as an entomologist in favor of "a diving holiday"; he became a marine biologist overnight, without having ever attended a lecture in the subject; and he became the Great Barrier Reef's first full-time scientific researcher.[16]

Joining James Cook University at the mid-Reef port of Townsville in November 1972 launched the final phase of Charlie's *Beagle*-like apprenticeship. JCU had been designated a university for only two years and was still a frontier institution. Its stiff new zoology professor, Welshman Cyril Burdon-Jones, John Busst's collaborator, was plainly unimpressed by the new postdoctoral fellow. Surveying the bearded twenty-seven-year-old in shorts and flip-flops who'd reported for duty, the professor raised his eyebrows, pointed vaguely toward the Reef, and gave Charlie his orders: "Your job is to go out there and do something," adding as an afterthought, "and try to stay out of trouble." Both instructions were to prove difficult.[17]

Charlie's first task was to identify and map the local corals, but the taxonomic volumes in the university library were technically forbidding, and tricky to match with what he actually saw underwater. Thanks to the new "sport" of scuba diving, he was entering a world that had been closed to most previous Reef researchers. Yet the results were seriously disconcerting, because the corals appeared to be far more plastic and protean than their confident taxonomic descriptions suggested.

"[T]he essential problem," Charlie discovered, "was that on the reef slopes, corals varied their growth forms and skeletal structures according to the environment in which they grew." A well-known type might appear stumpy and compact on the wave-pounded reef front, but as he swam down the protected slope it would mutate into delicate, long-branched fans. And "to make matters worse, big colonies regularly had one growth form on their top, different growth forms on their sides, and another form at their base." All too often the taxonomic guides described each of these variations as a different species. Charlie assumed the experts were correct, but he couldn't make sense of the bewildering variations. The fact that particular coral species usually clustered together on the same reef patches "suggested that there was some sort of order, or natural reality, behind the apparent chaos," but what was it?[18]

The only way out of this impasse was to improvise. Abandoning the formal taxonomic lexicon, Charlie began to identify familiar species with nicknames of his own. He then used a novel way of describing them that he'd come across in reading the work of rain forest ecologists like Len Webb and Geoff Tracey. Confronted with similarly chaotic environmental variations, they had mapped the dynamic relationships between different rain forest species and communities, looking for patterns of connection, struggle, and dominance—a process they called "population or community ecology." Instinctively, Charlie had also made his own transition to newly emerging ecological methods, including some similar to those improvised by biologists with Yonge's expedition to the Low Isles in 1928–29.[19]

Being free of the constricting traditional assumptions of professional taxonomists had its drawbacks, however, as Charlie discovered in June 1973 at the Second International Coral Reef Symposium. This was hosted by the Great Barrier Reef Committee on a ten-thousand-ton cruise ship, the *Marco Polo*, chartered to make a ten-day return voyage between Brisbane and Lizard Island in the far north, carrying 264 of the world's leading coral scientists. It was a brilliant idea: they drank, dived, danced, delivered papers, and argued while the sublime objects of their study slid past the bows.

Charlie's own paper, though, received a tough reception. Patricia Mather, secretary of the GBRC, icily rejected it from the proposed published proceedings of the talks because, as a scientist friend of Charlie's told him, "she doesn't approve of people like you messing about with taxonomy." Burdon-Jones underscored the point by interrupting Charlie on the dance floor to repeat what had become a stock refrain: that he intended replacing him with a proper postdoctoral researcher. Charlie had heard the threat once too often. He grabbed the professor by his tie and threatened to toss him overboard.[20]

Thankfully, the ship also carried some less stultified thinkers. One was Isobel Bennett, who wandered up to the shaggy young man she'd last seen as a small boy and welcomed him into the fold of coral science: "I thought you'd come home," she said warmly. David Stoddart, perhaps the most famous scientist on board, whose portly figure and Cambridge accent turned out to be no signifier of orthodoxy, told Charlie that his paper was "the best of the lot," and promised it would be published in spite of Mather's veto. More importantly, he asked Charlie to join the forthcoming northern leg of a Great Barrier Reef expedition he was currently leading for the Royal Society.

This dream opportunity was approved thanks to Ken Back, vice-chancellor of JCU and a man relaxed enough to share Charlie's "love of the sea." Back was prepared to put the university's research vessel, the *Kirby*, in the hands of a local skipper, Davey Duncan, who'd fished most of the coastlines and islands between Townsville and the Torres Strait. Duncan agreed to take on the considerable challenge of navigating through "3000 kilometers of poorly charted reef waters," and in November 1973 the *Kirby*, riding low in the water from its load of scuba equipment, diesel drums, food boxes, and beer cartons, departed Townsville for the Top End.[21]

The Stoddart expedition was Charlie's underwater counterpart to Darwin's daring *Beagle* voyage. Skirting Orpheus Island, they followed the ocean-crashed outer Barrier reefs up to the Torres Strait, with "no pre-determined route, nor pre-imagined outcome." At times they had to use Matthew Flinders's nineteenth-century charts because there were none better. Stoddart, who couldn't swim, concentrated on surveying the numerous coral cays, while Charlie dived, experiencing his "first real taste of the joy of discovery, of seeing and doing incredible things." Residual dreams of the Amazon vanished before the Reef's realities.[22]

One of their rough charts showed the 9,800-feet-deep Queensland Trough to be situated somewhere near the ribbon reef of Tijou, so they anchored the boat and moved around to the reef's boisterous outer face. Charlie dived down the steeply plunging sides until confronted by the "abyssal depths" of the trough. As the first diver ever to see this dizzying sight, he pushed downward through the clear water until he reached the limits of his body's capacity. Later, diving in Tijou's lagoon, he had to fend off two unusually frenzied sharks, though these proved to be only a vanguard of the seventeen others he counted circling feverishly there the following day.

Even farther north, just below the Torres Strait, they reached the approximately eighty-acre coral cay of Raine Island, which had rarely been explored by scientists since Joseph Beete Jukes landed there from the HMS *Fly* in 1844, in the company of a group of convicts sent to build a beacon. To the *Kirby* mariners it felt like a "re-discovery." Clamoring colonies of seabirds shared the sand and rocks with "the biggest turtle rookery in the world." Twenty thousand turtles scrabbled up the beaches to lay their eggs, while their slower brethren were being tossed in the air by giant tiger sharks.[23]

Like Jukes 130 years earlier, Charlie found it difficult to convey the primeval immensity of the Great Barrier Reef to people who'd not been there: "we just haven't got enough words." Wild, for example, was an overused term, but to Charlie it meant "never seeing any other sign of humanity, never seeing any evidence that there was anybody else [who'd] ever been on this earth, except at dusk, you would sometimes see a satellite go over, and that's it. You get this feeling of wilderness, of immense amounts of wilderness . . . the feeling of being an explorer, getting into places that no human being has ever seen before . . ."[24]

Occasionally, moved by some ancient urge to merge with the Reef's subterranean life, he would wait for a full moon and sneak out alone to the outer reefs. Plunging down into the blackness, he would drift to the sandy bottom and lie back, relaxing in a spectral world. "In very clear water the whole reef becomes a weird silver-gray and is full of life. Crustaceans crawl everywhere. Corals, seen mostly in silhouette have tentacles extended, making all manner of other-worldly shapes. Some fish are asleep, others are hunting. Sharks appear out of the silver, perhaps to circle around to check the newcomer, then to vanish. They are hunting, giving further spice to an already overloaded atmosphere. Reefs seen by moonlight, when one is all alone, are wondrous peaceful places."[25]

All apprenticeships must end some day. Darwin evaded post-*Beagle* restrictions on his intellectual freedom by becoming an independent scholar. Wisely he bought a house in Kent designed to meet every need. It became a family home, a refuge, a base for correspondence, a library for reading, an office for writing, a laboratory for experimenting, and a garden for experiencing nature.[26]

Charlie had no such luxury. As his postdoctoral scholarship drew to an end, he needed a paid job. Salvation, once again, came from the fairy-godmother figure of Isobel Bennett. Late one evening in 1973, in foul weather, he leaped off the boat after returning from a dive near Townsville, to see Issie emerging from the dark. She shone her torch on his face and announced that he had to apply for the position of foundation coral scientist at the new Australian Institute of Marine Science (AIMS) in Townsville—the fruits, indirectly, of John Busst's lobbying a decade earlier.

Charlie hedged. He was ineligible, he said. "I am a naturalist. Not a scientist. I'm a nature-lover naturalist, that's what I am." Nobody in those times paid naturalists to observe and think about what they saw.

The days of Darwin were long gone. AIMS would want a credentialed marine biologist. Waving away his protests, Issie warned him she would ring AIMS to check that he'd applied.[27]

In the end Charlie didn't need to apply. The newly appointed director of AIMS was an earthy, energetic American called Red Gilmartin, who'd been attracted to Townsville by the challenge of starting a first-class research center on the frontier among straight-talking Australians. And he found Charlie suitably blunt when he met him, at a dinner they'd been invited to by the vice-chancellor. Next morning Charlie picked up the phone to hear himself being offered a three-year job as the first AIMS scientist. Red wasted no time. Charlie was to start the following day in his new lab—a steel shed at Cape Pallarenda, a few miles northeast of Townsville. His brief was to write a series of monographs on the corals of the Great Barrier Reef.

Charlie now had no choice but to make himself a genuine taxonomist. Red told him bluntly that a comprehensive coral taxonomy had to precede any "meaningful ecological work." Fortunately, Red believed in supporting his missions properly, and at least in the early days there were few bureaucrats to stand in the way of Charlie's curiosity. An old diving colleague from his Solitary Islands days was appointed as his assistant, and gradually Charlie acquired an impressive collection of field scientists to collaborate on the monumental task.[28]

First he needed to visit the great natural-history museums of London, Paris, Berlin, and Washington to examine the "type specimens" that underpinned existing coral taxonomy. To his chagrin, many of these foundational types proved flawed. Deskbound taxonomists of the past had named new species with promiscuous abandon, when these were often only local variations of the same species. Charlie later estimated that five thousand different names had been given to around two hundred coral species. A major offender was none other than the legendary British pioneer of coral taxonomy, Henry Bernard. Had the great man spent just one day in the Great Barrier Reef, Charlie reflected wryly, "his world would have been turned upside down."[29]

For the next eight years Charlie and his colleagues threw themselves into a grinding routine of identifying, observing, mapping, collecting, illustrating, photographing, and describing the myriad variations of live reef-growing coral colonies (of the order Scleractinia), from the southern temperate to the far northern tropical zones of the Barrier. In the process, Charlie clocked up an astonishing seven thousand hours of diving.

Scleractinia of Eastern Australia, the first in what would become his multivolume monograph series, appeared in 1976 to a rough reception from Northern Hemisphere paleontologists. Several wrote sniffy reviews or scathing letters, taking exception to Charlie's unorthodox views on environmental variations within species. He took some consolation, however, from the fact that one of the world's most distinguished taxonomists, John W. Wells, professor of geology at Cornell University, thought the book "refreshingly original."

Even so, Wells, who was a specialist on Marshall Islands corals, suggested that Charlie had exaggerated the problem of environmental variations in that particular location. There remained only one way for Charlie to prove his point to the famous but nondiving professor. Visiting a Marshall Islands reef with Wells, he swam down around 165 feet and, at decreasing depths, collected samples of a single well-known coral species that Wells had described. Laying these out on a bench, Charlie left the taxonomist to examine the samples at length.

Shaking his head, Wells eventually conceded that they were undoubtedly a single species, but that paleontologists "would have called most of these specimens a different species and . . . would probably have made several genera of them." In fact, so converted was he to Charlie's viewpoint that he now expressed doubts that a single taxonomic framework of the world's corals would ever be possible.[30]

After several extended visits to Western Australia in the early 1980s, Charlie began to fear that Wells was right: the chaos in his mind seemed to be increasing, if anything. The Indian Ocean reef of Ningaloo was fascinating but additionally puzzling. After ten minutes of snorkeling there, Charlie realized that the ocean currents bordering Africa, South America, and Australia could not run in a northerly direction up Western Australia's coast, as was commonly thought. Ningaloo was rich with distinctive Indonesian corals that must have been carried by southward currents. Moreover, at the more southerly reefs of the Houtman Abrolhos chain of islands, where tropical corals battled with cold-water algae, he found a mixture of corals like nowhere else. Yet there was something slightly odd about many of the species there compared to their counterparts on the Barrier Reef. Were his previous identifications of Scleractinia somehow flawed?[31]

For several decades Charlie pursued this nagging puzzle of the divergences between the same species at different locations. His quest took him to hundreds of reefs in both hemispheres and across the vast Indian and Pacific oceans. He dived and collected in Japan, the Philippines, Indonesia, the Cocos (Keeling) Islands, and then farther afield, in Zanzibar, off the east coast of Africa, and at remote Clipperton Atoll in the eastern Pacific, about 750 miles southwest of Mexico. Always he traveled by boat, always he worked with locals, and always he spent hours underwater, observing and memorizing.

The many comparisons he made led to a fresh discovery. When his records of species distribution and diversity were mapped, gathered, and interpreted—material that later became Charlie's computer database, Coral Geographic—they confirmed that the Great Barrier Reef's coral diversity, rich though it was, didn't match that of Indonesia and the Philippines. It was clear that the greatest coral diversity in the world was centered on a roughly triangular area within the Central Indo-Pacific, known ever since as "The Coral Triangle."[32]

Charlie Veron exploring coral species variations on Emre Turak Reef
in Madagascar (Courtesy of Charlie Veron)

The more Charlie ventured into these newly discovered regions, the less certain he became about his previous Barrier Reef observations: "My confidence faltered, not because the corals were different, but because most were neither different nor the same." A trip to the reefs of Vanuatu, which were linked to Australian reefs by ocean currents sweeping through the Coral Sea, only intensified his anguish. When Vanuatu reefs featured species in common with the Barrier Reef, they proved "virtually identical," yet not actually so. Despairing, Charlie teetered on the edge of giving up his twenty-year struggle to produce a unified taxonomy of coral species. "I can't publish work on something that's only half right," he tortured himself.[33]

> . . . a well-defined species on the central Great Barrier Reef might be a bit different at Vanuatu or the West Australian coast and be a bit different in other ways in Indonesia. In Fiji it might be a different species altogether but if it turned up in Papua New Guinea it might be seen to be a hybrid. These sorts of patterns recurred in one country after another and did so for most species. The more I worked in different countries, the deeper the problem became. It was an unhappy state of affairs because there was no way through; more work made the problem worse not better. There was no solution.[34]

One morning Charlie got up around 5:00 a.m. to begin work, and as usual made himself a cup of coffee. As the kettle boiled he had a flash of insight that gave him an answer. There was nothing mystical about this moment of intuition: he'd always believed that humans were "good at subconsciously synthesizing information, and that many ideas simply come of their own volition rather than as the intended outcome of planned research."

At the same time, Charlie reflected, most humans were not so good at grasping aspects of nature that couldn't be clearly defined or placed into hierarchies, even though nature's products were "seldom organized into species at all." Now he saw that, considered over vast geographical space and long swathes of geological time, coral species were malleable and temporary units, fluidly interlinked by their genes to other units, and forming ever-changing patterns. Corals had to be treated as continua, not as fixed, isolated units.[35]

If corals did become isolated and unable to interbreed for long periods, Charlie ruminated, then the fittest among them would, as Darwin had argued, be selected to survive and so eventually form a genetically cohesive new species. But the breeding of corals was one of those simple natural processes that produce inordinately complex results. While corals struggle with one another to survive and dominate within a particular area, the spectacular processes of mass spawning by which they breed hurls out gene-carrying larvae into the ocean, to be carried potentially vast distances by the chance actions of currents. Eventually the larvae might settle to grow and breed within a distant coral empire. Here they would, over geological spans of time, intermix to produce new variations, reconnections between former variations, or even "fuzzy" hybrids.

Realizing that in evolutionary terms corals behave a lot like plants, Charlie consulted a colleague at AIMS, who explained how his insight aligned with the concept of "reticulate evolution," a process that plant geneticists, in particular, consider a convincing alternative explanation of how species change over space and time. The theory of reticulate evolution argues for the formation of a netlike skein of evolutionary threads, determined by environment, rather than the famous branching tree of natural selection sketched by Darwin in the *Origin*. Had Darwin known what modern geneticists know, he might well have agreed, for he was, in Charlie's words, "a fabulous thinker, just a wonderful, wonderful thinker." The irony was that Darwin, in his day, acquired a reputation as a heretic for denying the divine fixity of species, and now Charlie was at risk of making himself a heretic by dismantling the taxonomies of species altogether.[36]

Like his namesake, Charlie knew he would have to do a hellish amount of reading and research to make his controversial case. Beginning in 1992, he navigated a maze of different disciplines, from paleontology, taxonomy, biological oceanography, and ecology to systematics and molecular science. Darwin had been forced to do something similar, but disciplinary boundaries were neither as numerous nor as heavily patrolled in his day. Specialized traditions and terminologies were now often impenetrable to outsiders, an exasperating problem for Charlie because he wanted to simplify them in order to reach a general readership.

Mary Stafford-Smith, a colleague, coral scientist, and gifted editor, gave him vital help. She had no compunction about sending him grumbling back to the writing desk, from which he would usually emerge with a clarified text and an improved humor. The resulting book, *Corals in Space and Time* (1995), was revolutionary by any standard, for it contained the dual discoveries of the "Coral Triangle" and "Reticulate Evolution in the Scleractinia." The magisterial journal *Science* devoted a full article to its contents, and Charlie was in 2004 awarded the prestigious Darwin Medal by the International Society for Reef Studies.[37]

Yet even now Charlie feels that the full acceptance of reticulate evolution within the scientific world has been slow. More than a decade on, young marine scholars refer to it as "the final frontier," rather than a proven approach for understanding the evolution of corals. The high degree of difficulty of working with genetic continua and the challenge this poses to some versions of natural selection make for strong forces of resistance.[38]

Charlie was conscious that the seeds of a different and graver problem also shadowed *Corals in Space and Time*. His realization of this new conundrum had personal as well as intellectual roots. In the midst of his long, testing labors on coral evolution, a family tragedy drove him to think intently about mortality and survival. Just as Charles Darwin, struggling to finalize his theory of evolution, had been shaken by emotional loss, illness, and domestic strain, so it was with Charlie Veron.

Charlie's tropical equivalent of Down House—Darwin's residence in the peaceful hamlet of Downe—was Rivendell, named after the elves' abode in Tolkien's *Lord of the Rings*: "a perfect house, whether you like food or sleep or story-telling, or just sitting and thinking best." Charlie's Rivendell was situated on a five-acre bush block with tree-edged river frontage, twenty minutes' drive from Townsville. He and Kirsty had bought the place in 1976, and their effervescent little daughter, Noni, had hammered a stake into the ground thirty-two feet from the riverbank to mark the house's future entrance. It was an ideal environment for a free-spirited, dreamy, nature-loving child like Noni. Charlie built her a tree house in the branches of a large eucalyptus that overhung the river. They called it Gum Leaf, and Noni, at the age of six, resolutely insisted on sleeping there alone on the first night. For her, Rivendell's relative isolation was compensated for by the pleasures of books, dogs, horses, a piano, and, after 1978, a little sister, Katie.[39]

Yet much of Charlie and Kirsty's domestic life had proved grindingly tough, just like it had been for the Darwins. Among other problems, Katie's infancy was haunted by acute respiratory and feeding problems. For eight months, as her life guttered, she had to be watched continuously. For six of those months Charlie and Kirsty slept only on alternate nights. Inevitably, Noni felt the brunt of her parents' distraction and exhaustion, and began to show bursts of discontent at home and school. Yet "always [she] would respond to reason and always she would accept what was just."[40]

But nature knows nothing of justice. In April 1980 Charlie was working in Hong Kong when he received a phone call from Kirsty to say that Noni had drowned while playing in a creek with a friend. A light went out in Charlie's life. Somehow he got himself back to Townsville the next morning to see Noni lying in her coffin. "I kissed her face, it was frozen. This is the worst memory of my life." She was cremated on her tenth birthday.[41]

One hundred and twenty-nine years earlier, almost to the day, Charles and Emma Darwin's ten-year-old daughter Annie also died. Desperate to keep her with him, Darwin wrote: "Her dear face now rises before me, as she used sometimes to come running downstairs with a stolen pinch of snuff for me, her whole form radiant with the pleasure of giving." With tears streaming down his face, he concluded his short memoir of Annie in utter desolation: "We have lost the joy of the household, and the solace of our old age. She must have known how we loved her. Oh, that she could now know how deeply, how tenderly, we do still and shall ever love her dear joyous face."[42]

Similarly stricken, Charlie Veron survived in a haze of sleeping pills, and Rivendell became a refuge for memories. He was frantic to keep those memories alive: "I was always talking to Noni, even when I was actually talking to someone else." Over time, though, these vivid conversations grew muffled and Noni's voice began to fade. For Charlie, this was "almost a second dying." As always in his life, he found consolation in the redemptive energies of nature. The spot where Noni had kept her ducks and geese became a memory garden filled with plants and trees donated by friends. In it was one special tree: "at the edge of the garden is a tall umbrella tree, the last present that Noni ever gave to me. It was for my birthday, and was decorated with coins stuck to the leaves with sticky tape to show me that 'money grows on trees.'"[43]

Weighed down by "such unrelentingly bad times," life for Charlie and his wife dragged, and although they remained close and supportive of each other, they eventually agreed to divorce. "I think the death of a child is the biggest thing someone can live through, it takes away almost everything," Charlie later told a friend. And it is surely true, too, that when you have faced the death of someone inconsolably dear, nothing else can defeat you. Charlie survived this dark night of the soul thanks to Rivendell, the solace of diving on the Reef, and his much loved dogs. Sometimes he simply took the phone off the hook and lived the life of a recluse and an automaton. Looking back, he sees himself then as "a cot case."[44]

He was redeemed by the chance of a second romance, with Mary Stafford-Smith, the scientist who edited *Corals in Space and Time*. Mary was working on Lizard and Orpheus islands, on the effects of sediment on corals, and became the companion of his heart as well as his mind. She left behind her birth family in England to revitalize Charlie and Rivendell with her own energies and values. As well as being a stepmother to Katie, she and Charlie started a new family. Two young children brought life, laughter, and love back to the house on the river.

Charlie's intense personal reminder of the contingencies and fragilities of life found echoes in his research. Writing *Corals in Space and Time*, he was forced to investigate the fate of the world's corals in the past and present. He studied paleontological analyses of previous reef extinctions and accrued more and more evidence of the effects, on the Great Barrier Reef in particular, of changing sea levels, temperature stresses, predation by crown-of-thorns starfish, and human-influenced changes in nutrient levels. All this sharpened his long-gestating concern about the health of the Great Barrier and other world reefs.

In the aftermath of the book's publication, he and Mary began discussing the idea of a glossy, coffee-table book about world corals for a general audience, "not just to produce another book, but to open the eyes of the world to what was emerging as an urgent need to conserve corals." It was the crystallization of a new joint mission: "to win some hearts as well as minds."[45]

The project gathered enormous momentum with the enthusiastic support of like-minded scientists. Around seventy underwater photographers gave their work for free, and illustrator Geoff Kelly produced a series of exquisite drawings and paintings. Charlie supplied most of the encyclopedic thumbnail analyses and took overall control as general editor, while Mary edited the science and produced the book. AIMS covered the printing, and in October 2000 the three-volume *Corals of the World* was launched to great acclaim at the International Coral Reef Symposium in Bali, where its message of reef fragility and degradation added to a rising global alarm.[46]

An instinctive conservationist, Charlie had been troubled way back in the 1970s by the extent of the damage caused by coral-eating crown-of-thorns starfish. He'd become convinced that numbers of the latter were soaring due to overfishing of the starfish's natural predators, and that survival of the millions of larvae expelled annually into the ocean currents was enhanced by the growing levels of chemical pollution. (Crown-of-thorns larvae thrive in polluted waters.) What provoked him to fury, though, was the way in which the vested interests of tourist developers and politicians, combined with the craven behavior of government bureaucracies, worked deliberately to discourage scientists from studying the problem. It was the onset of a process, ubiquitous today, whereby scientists were no longer free to pick their own questions or seek their own answers. Politicians of various kinds set self-interested agendas based on money and voter appeal, instructing bureaucrats to herd the scientists into yards like cattle. Concerned scientists were purposely deflected from working on the crown-of-thorns starfish problem, even though it had in no way receded.[47]

Looking back, Charlie realized that like most of his generation he'd taken for granted that "the oceans [were] limitless and the marine world indestructible," including the vast, relatively well-managed region of the Great Barrier Reef Marine Park. The fact that the Central Indo-Pacific functioned as the prime disperser of coral biodiversity had always been worrying because of the region's lack of legal protection. Diver friends had long urged him to visit the spectacular reefs of eastern Indonesia, but by the time he got there in the early 1990s it was too late. Reefs that had run for thousands of miles were now masses of rubble. He found one 980-foot stretch of breathtaking richness and beauty that had survived only because a local tourist operator was having it patrolled by a guard with a gun.

Two weeks after Charlie photographed this reef, its lone guard fell ill and local fishermen moved in with bombs made from diesel and garden fertilizer to explode swathes of coral and scoop up the dead fish for a meal or two. Any corals that survived the bombs would die anyway, because, as fishermen well know, coral reefs need fish to keep algae under control: "if you removed fish, within months the reef would be smothered with algae." There were few, if any, samples of sharks and large fish among the corpses; they'd been fished out long ago.[48]

Charlie had seen his first patch of coral bleaching off Palm Island in the early 1980s, a tiny, four-inch clump of white skeleton that he photographed as a curio. "And then I saw a whammy, a mass bleaching event . . . where everything turns white and dies. Sometimes it's only the fast-growing branching corals, but some of the others are . . . horrible to see—corals that are four, five, six hundred years old, they die too. It's a very recent thing." The first recorded global mass bleaching occurred in 1981–82. At the site where Charlie was then working, a beautiful embayment off Orpheus Island, seventy miles north of Townsville, 60 to 80 percent of the corals bleached. Within the Reef region overall, around 66 percent of inshore and 14 percent of offshore reefs registered moderate to high levels of damage.[49]

The next major spate of mass bleaching, in 1997–98, hammered reefs in more than fifty countries, through the Indo-Pacific, the Red Sea, the Caribbean, and even among the hot-water corals of the Arabian Sea. On the Great Barrier Reef the bleaching coincided with the warmest sea temperatures ever recorded. In an even worse mass bleaching event in 2001–02, the global damage also confirmed a close connection with El Niño weather cycles. Catastrophic global warming had arrived.[50]

Up until the 1980s, theories that the earth's climate was changing because of human-engendered greenhouse gases such as carbon dioxide and methane seemed remote from Charlie's concerns as a reef scientist. Like any properly cautious researcher, he was initially wary of some of the startling claims of coral scientists—those of his friend Ove Hoegh-Guldberg, for example. But the more Charlie immersed himself in the torrent of papers and data on the subject coming from all over the world, the more convinced he became that Ove was right, and that corals were "the canaries of climate change." In the same way that the deaths of those tiny yellow birds had alerted nineteenth-century miners to the presence of poisonous gases, so reef-growing corals, which seemed peculiarly susceptible to increases in heat and light, were now alerting scientists to climatic changes.

Charlie's research told him that during El Niño weather cycles the surface seawaters in the Reef lagoon, already heated to unusually high levels by greenhouse-gas-induced warming, were being pulsed from the Western Pacific Warm Pool onto the Barrier Reef's delicate living corals. When corals are exposed to temperatures two or three degrees hotter than their evolved maximum of eighty-eight degrees Fahrenheit, along with increased levels of sunlight, it's lethal. The powerhouse algae that live in the corals' tissues, providing their color and food through photosynthesis, begin to pump out oxygen at levels toxic to their polyp hosts. The corals must expel their symbiotic life supports or die. Row upon row of stark white skeletons are the result.

These damaged corals are capable of regeneration if water temperatures return to normal and water quality remains good, but the frequency and intensity of bleaching outbreaks is now such that the percentage of reef loss from coral deaths will increase dramatically. Charlie predicts that the widening and deepening of the Western Pacific Warm Pool through climate change will mean that "every year will effectively become an El Niño year as far as the corals are concerned."[51]

It's Charlie's hope that some as yet unknown strains of symbiotic algae, better able to cope with a heat-stressed world, might eventually form new partnerships with the corals. Or that the adaptive energies of fast-growing corals like *Acropora* might somehow outpace the rate of bleaching. Or that pockets of coral lying in shadowed refuges on cool, deep reef slopes, or in recently discovered deep waters, might survive to become agents of future renewal. But heat is not the only problem corals face.[52]

Charlie's multidisciplinary investigations of the significance and impact of reef extinctions over time show that other destructive synergies are already in motion, and may be impossible to stop. Reefs, he points out, are nature's archives and historians. They are complex data banks that record evidence of environmental changes from millions of years ago up to the present. Imprinted in fossil typography are the stories of the mass-extinction events of the geological past, including their likely causes. These archives tell us that four out of the five previous mass extinctions of coral reefs on our planet were linked to the carbon cycle. They were caused by changes to the ocean's chemistry brought about by absorption of carbon dioxide and methane, through a process of "acidification."[53]

Today's culprits are the same gases—carbon dioxide and methane— though their increased presence is not due to the massive meteor strikes or volcanic eruptions that caused earlier catastrophes. We humans are doing that work, knowingly pumping these gases into the atmosphere at unprecedented rates. Already the oceans, the planet's usual absorber of these gases, have reached a third of their capacity to soak them up and balance them chemically. Stealthily, the oceans of the world have already begun the process that scientists call "commitment," which in this case refers to the "unstoppable inevitability" of acidification that presages destruction long before it is clearly visible. Eventually—possibly as early as 2050—we will have reached the point where coral skeletons become soluble in seawater. Carbonate rock, including reefs, will start dissolving, like "a giant antacid tablet," as Charlie describes it.[54]

If, as AIMS tells us, the Great Barrier has already lost half its coral cover during the last twenty-seven years through bleaching, cyclones, pollution, and crown-of-thorns starfish, what will happen to this figure as the effects of acidification take hold? Reef corals will be among the first organisms in the oceans to be affected by this alarming process, stricken, in effect, with a fatal form of "coralline osteoporosis." Their aragonite skeletons will either stop growing altogether or become too brittle to resist the eroding effects of waves. Needless to say, they will crumble more dramatically before the lashing cyclonic events that are now on the increase. Rising sea levels caused by the melting of the ice caps will probably also compound the inability of corals to grow, by diminishing their exposure to light.[55]

Phytoplankton, the food of tiny krill, a key element in the food web of the southern oceans, will be equally affected by acidification. And who knows what terrible chain of ecological consequences will follow? We could face something like the great K-T extinction at the end of the Cretaceous period, around sixty-five million years ago, a disaster that put an end to most life on our planet for millions of years. Long before this, we Australians, at least, will "see a meltdown in coastal economies with devastating cost to natural environments and human societies." But eventually a remorseless domino effect, beginning with coral reefs and their marine communities, will presage a succession of ecosystem disasters. The earth's sixth mass extinction event will have arrived.[56]

So, Charlie Veron, a man who has lived and worked on the Great Barrier Reef for most of his life, finds himself in the agonizing position of having to be a prophet of its extinction. We can't wonder that he feels "very very sad. It's real, day in, day out, and I work on this, day in, day out. It's like seeing a house on fire in slow motion . . . there's a fire to end all fires, and you're watching it in slow motion, and you have been for years."[57]

I know of few more poignant sights than the closing moments of that speech Charlie gave in July 2009 to the hushed room of scientists and citizens at the Royal Society. Tossing aside his notes he apologizes to the audience in a strained, faltering voice for having delivered such a miserable talk. He urges his listeners to think about what they've heard. The implications of what he's revealed do, literally, "beggar belief."

"Use your influence," he pleads. "For the future of the planet, help to get this story recognized. It is not a fairy tale: it is reality."[58]

EPILOGUE

A Country of the Heart

THE DEATH OF THE REEF and the onset of a new epoch of mass extinction—it's a chilling message, and a depressing note on which to end. I share Charlie Veron's dilemma about this. As a scientist, he's told the story of the Reef's past, present, and future as truthfully as he can, in the hope that people will confront its likely demise. At the same time he worries, as I do, that the grimness of this will provoke fatalism, despair. I've heard even well-meaning people say they're sick of misery stories about climate change, and if the death of the Reef is inevitable, what's the point of worrying? Why not get on with living and just enjoy the place while we can?

Oddly, though, I find myself more optimistic now than I was a dozen years ago, when my voyage on the *Endeavour* replica triggered the relationship with the Reef that led to the writing of this book. It's true that from a scientific viewpoint things look bleak. Corals are indeed the canaries of climate change, and they face death from many more threats than noxious gases in coalmines.

Yet nothing is certain when it comes to predicting the operations of nature, which has the capacity to defy as well as exceed our worst prognostications. Perhaps in the future, as Charlie and other scientists have speculated, some corals and symbiotic algae will adapt to the massive changes currently occurring in their environment; perhaps the recent discovery of corals growing in near darkness at great depths will provide a nucleus of resistance. It's also possible that scientists and geoengineers will find ingenious ways of slowing or reversing the toxic trend of greenhouse gases.

But what has buoyed me beyond mere dogged hope is the spirit of optimism I've met among the peoples of the Reef while working on this book. In late 2012 I returned there yet again, this time to film some of the sites I'd been writing about. With me were Mike Bluett, a friend and a brilliant documentary filmmaker; marine biologist Dean Miller, also a splendid cameraman and, like Bluett, from Cairns; and my son Andrew McCalman, who happens to be a talented stills and video photographer.

Our first stop, fittingly, was Cooktown. My 2001 *Endeavour* visit had been a miserable failure: the moving footage of clan elder Eric Deeral reflecting on the meaning of Cook's encounter for his people today was subsequently cut from all three TV versions. I'd also been aware back then that every year the white inhabitants of Cooktown showed a similar disregard for Aboriginal opinion when they reenacted the *Endeavour*'s visit at their Discovery Festival. Pulitzer Prize–winning author Tony Horwitz had been in Cooktown only a few months before me in 2001, and he witnessed one such festival, which his white guide promised him would be "a shocking piss-up." The farcical costume restaging was accompanied by "a wet T-shirt contest . . . [and] a race down the main street by revelers pulling beer coolers mounted on wheels." A local policeman assured the dazed Horwitz that this particular "three-day bash" had been unusually sedate: only a few years earlier it had generated ninety-seven arrests for drunkenness.[1]

But much had changed when we returned a dozen years later. Although Eric Deeral was too frail and ill to see us, his legacy was everywhere apparent. Years of patient work as an MP and elder had transformed the social landscape. Much was also owed to Eric's extraordinary niece, Alberta Hornsby (née Gibson). Alberta is a phenomenon: a small woman with a gentle voice that belies a granitic determination. Listening to her uncle's quiet wisdom, she realized that beneath the one-dimensional story of the encounter between blacks and whites were complex layers of historical and contemporary meaning, within which lay the seeds of a renewal of pride and respect for today's Guugu Yimithirr and Kuuku Yulanji.[2]

So Alberta became a historian, and a myth buster for black and white alike. Her reinterpretation of the encounter came not just from scouring books and documents, but also from patient talks with elders, and discussions with scholars at conferences and seminars across the country.

As we filmed, Alberta and I sat on the ground at Grassy Hill, the stony rise that overlooks Cooktown. Below us were the snakelike folds of the Endeavour River and the distant waters spotted with shallow green smears of coral. This was the spot from which Alberta's ancestors had first spied the three-masted *Endeavour* limping through the shoals to the riverbank. I asked her about the aspects of the encounter that puzzled me, in particular the sequence that led the Guugu Yimithirr to first ignore, then fight, and ultimately make peace with their invaders. Her informed explanation of the mutual bewilderment of the Europeans and the Bama has, I hope, deepened my own account. The Bama, she said, regard the encounter as a tragedy of misunderstanding, one that yet contains potential for change. Eric Deeral's knowledge and insight, for example, are now incorporated in Cooktown's annual reenactment. Since 2010 the Discovery Festival has gone from being a white bacchanal to a genuine cross-cultural event. The local policeman must be bemused.

By the time we waded onto Dunk Island for our next bout of filming, gale-force winds were driving sheets of rain along Brammo Beach. This seemed appropriate, for Ted Banfield's paradise isle now looked like a First World War battlefield. The luxury resort had been devastated over two days in February 2011 by Cyclone Yasi, which had ripped the fronts off apartments, peeled back roofs, twisted the steel frames of the function hall, and dumped tons of sand into the swimming pool. Palm trees had been stripped to stalks, and the nearby forest canopy decapitated. Yasi had made a mockery of Hideaway Resorts' "cyclone-rated buildings to category five," themselves a response to Cyclone Larry, which had inflicted twenty million Australian dollars' worth of damage on the same site only four years earlier.

We were given generous access to the surviving amenities by the thoughtful general manager of the resort and his remnant staff, who were still struggling to bring order to the chaos. The site was up for sale, but who, we wondered, would buy a place so firmly on the flight path of successive and ever more ferocious cyclones.

We'd come to meet Susi Kirk, who was helping out at the resort but lived apart from it, in a storm-battered mudbrick house. She'd arrived in the early 1970s, a twentysomething swept up in the same romantic currents that had carried John Busst and other "escape artists" to the beautiful Family group of southern Reef islands. In 1974 the charismatic Bruce Arthur, who'd once been an Olympic wrestler and was now a tapestry maker, began an artists' colony on Dunk. It found buyers for its work among the resort's customers and added a splash of bohemia to the island's natural attractions. Bruce died in 1989, leaving Susi as guardian of his and his predecessors' legacies.

Susi roared up on an ancient motorbike with a chainsaw, crowbar, and several muddy shovels in a trailer. She exuded enthusiasm and practicality. She was currently caring for most of the island's abandoned livestock, clearing branch-clogged pathways, organizing transport for workers who wanted to escape to the mainland, and providing homemade cookies for those who stayed behind.

Wherever Susi went, you could feel morale lifting. She was passionate about the continuing spirit of Ted Banfield, and guided us by boat to one of his hidden places, the enchanting Cave of Falling Stars. Her faith in the island's natural resilience echoed a moment in 1918, when Bertha Banfield brushed aside Ted's cyclone-shattered pessimism by showing him fresh buds growing from a torn tree stump. When I reminded Susi of this she told a story of her own, about putting so much cement on a crumbling mudbrick dome left to her by Bruce Arthur that she'd unwittingly turned it into a bunker. Providentially, this had afforded her protection from both Larry and Yasi: "And it will cope with the next one just as well."

After we'd left, we heard from the manager that Susi had spoken so passionately to a visiting New Zealand family that they'd decided to buy the resort, and rebuild it in a smaller, eco-friendly and more cyclone-resistant form. I was again reminded of Bertha's rallying words to Ted in 1918: "So let's get the clocks going. There is work to be done."[3]

The northern Cape York town of Lockhart River, the last of our filmsites, was no reputed paradise. Home to an Aboriginal community of seven to eight hundred people from five different language groups, it is cut off from road access during the wet season, and its inhabitants have a life expectancy twenty years lower than the Australian average.

Among these inhabitants are the descendants of the Uutaalnganu-speaking people who rescued the cabin boy Narcisse Pelletier and transformed him into Anco of the Sandbeach country. Like so many Aboriginal clans, this one, too, had been severed from its heartland, its members forcibly intermingled with others when the Anglican mission was disbanded around 1968. The remnant mission community was moved inland, to what has now become the Lockhart River township, on land adjacent to an old mining airstrip.

Once a byword for drunkenness and violence, the town had, since 2001, become a dry community, with penalties for the sale, acquisition, or consumption of alcohol. Its artists, especially a group known as the Lockhart River Art Gang, were prominent in the extraordinary renaissance of Aboriginal art. And the community's mayor, Wayne Butcher, a dynamic man of Uutaalnganu descent, had masterminded a drive to restore the ancient coastal culture of the Pama Malngkana, or Sandbeach people.

In Anco's time the Pama Malngkana were brilliant hunters and fishers in the Barrier Reef lagoon, and later that century they provided the trepang and trochus-shell industries with skilled luggermen. But resource exhaustion and the supplanting of pearl-shell buttons by plastics put an end to all this. When founding a community-based fishing industry, Wayne and his colleagues had thus to start from scratch. They had to find new and sustainable produce, which they did in the form of rock lobsters and mud crabs, and also master new methods of catching, preserving, and transporting the produce to distant markets.

It was a tough undertaking, with numerous early setbacks, but Wayne was sustained as much by its social effects as its profits: "They are a different bunch of men when you take them out . . . [o]n a few days camping, chasing crays, their work ethics change, they change personally."[4]

One other vein of cultural renewal at Lockhart River took us by surprise. We checked into the modest timber units that are the town's visitor accommodation, and there on a stand in the foyer stood a well-thumbed copy of Stephanie Anderson's wonderful translation of Anco's story, *Pelletier: The Forgotten Castaway of Cape York*. Books were not much in evidence at Lockhart River, but we were to see well-used copies of this one at three separate localities.

Further confirmation of the story's impact came early the following morning, when Wayne was forced to put a quota on the number of locals joining us for a filmed beachside discussion of Anco. In the end we were rationed to just three; otherwise, as Wayne pointed out, Lockhart River local government would cease to function.

Dora, Gabriel, and Patrick were Uutaalnganu descendants who identified deeply with Anco's story. Gabriel's late grandfather, Alick Naiga, whose photograph appears in Anderson's book, passed on his knowledge of Sandbeach times to the anthropologist Athol Chase, who wrote a chapter on the Pama Malngkana for the new edition. Grandfather Alick had displayed a direct link to Anco in the form of identical initiation scarifications on his chest.

Dora, too, knew the details of Anco's rescue intimately. Through her we came to realize just how vital this story has become for today's Lockhart River peoples, offering as it does a portal into a world they thought they had lost: a rich, complex way of life that a combination of mission doctrine and government hostility sought to eradicate. Stephanie's translation of the book has revived memories of heartland, kin, lore, and language, and is indeed a source of hope.

Of course, I don't pretend that these three uplifting episodes can stand for the future of the Reef as a whole, but they do symbolize for me the resilience of the human heart, something I've encountered so often in writing this book. And ultimately we need both heart and mind if we are to meet the challenges that confront this unique country of sea, island, and coral that we call the Great Barrier Reef.

This brings me back to Charlie Veron, with whom this epilogue began, for nobody better embodies these twin properties. No single individual has explored the Reef more widely, no mind has engaged more trenchantly with its long history, and no heart has felt its present and future plight more intensely.

And in fact, nature offers us a model in this regard: the magical, reef-creating symbiosis between microscopic algae and a tiny polyp. We might see the algae as the heart that generates the energy its partner depends upon, and the polyp as the mind, having a purposive direction to build their joint production of coral. And also to protect it in the face of torrential forces of destruction—breakers, coral-feeders, cyclones, sediment, pollution, mining, overfishing, and water that is too hot and too acidic to bear.

It is a symbiosis which, as we have seen, has survived for some 240 million years, but which will split should those harsh forces so dictate. If anything can inspire us to prevent this, it's that very partnership itself, between two of the tiniest and most fragile creatures in the sea.

NOTES

PROLOGUE: A COUNTRY OF THE MIND

1. R. M. Ballantyne, *The Coral Island,* London, Thames Publishing, n.d. [1858], pp. 28–29, 34–35.
2. The Marine Park Authority consists of a chairman and a small executive committee of four or five members, chosen by the federal government minister with responsibility for the environment, and appointed by the governor general. Their mandate is to provide expert advice to the government on practical and policy matters relating to the management of the Barrier Reef Marine Park.
3. Quoted in B. Gammage, *The Biggest Estate on Earth: How Aborigines Made Australia,* Sydney, Allen and Unwin, 2011, p. 143.
4. D. Smyth, in *Aboriginal Maritime Culture in the Far Northern Section of the Great Barrier Reef Marine Park,* Townsville, Great Barrier Reef Marine Park Authority, 1995, pp. 9, 18–34, points out that marine resources are "part of a continuum of Aboriginal culture that binds the life of humans and animals, earth and sea, past and present." They provide identity as well as sustenance, in other words.
5. I. McCalman, MSS journal in possession of the author. See also, S. Baker's gracious survey: *The Ship: Retracing Cook's Endeavour Voyage,* London, BBC, 2002, p. 138. For some more trenchant accounts, see V. Agnew and J. Lamb, eds., *Criticism: A Quarterly for Literature and the Arts,* Special Issue: Extreme and Sentimental History, vol. 46, no. 3, Summer, 2004.
6. R. Macfarlane, *Mountains of the Mind: A History of a Fascination,* London, Granta, 2008, pp. 19–20. I am grateful to Ben Ball for drawing my attention to this marvelous book.

I. LABYRINTH: CAPTAIN COOK'S ENTRAPMENT

1. J. Cook, *The Journals of Captain James Cook on His Voyages of Discovery,* vol. 1, edited by J. C. Beaglehole, Cambridge, published for the Hakluyt Society at the University Press, 1955, pp. 320–32. See also the popular edition from which most readers came to know this story: J. Cook, *An Account of a Voyage Round the World with a Full Account of the*

Voyage of the Endeavour in the Year MDCCLXX Along the East Coast of Australia, Brisbane, edited by J. Hawkesworth, Smith and Paterson, 1969, pp. 517–18.

2. J. and M. Bowen, *The Great Barrier Reef: History, Science, Heritage,* New York, Cambridge University Press, 2002, p. 48.

3. J. Hamilton-Paterson, *Seven Tenths: The Sea and Its Thresholds,* New York, Europa Editions, 2009, pp. 115–20.

4. Cook, *Journals,* p. 342.

5. J. Gascoigne, *Captain Cook: Voyager Between Worlds,* London, Hambledon, Continuum, 2007, pp. 21–24.

6. J. Banks, *The Endeavour Journal of Joseph Banks,* 2 vols., edited by J. C. Beaglehole, Sydney, Trustees of Public Library of NSW in association with Angus and Robertson, 1963, vol. 2, pp. 66, 70–72.

7. I was shown this remarkable rock painting by senior custodians at Yarrabah when visiting there on the *Endeavour* replica in 2001.

8. Banks, *Journal,* vol. 2, p. 77.

9. Cook, *Journals,* p. 343.

10. Hawkesworth, *Account,* p. 544.

11. Cook, *Journals,* p. 344.

12. Ibid., pp. 345–47; Hawkesworth, *Account,* pp. 548–50, 552.

13. Hawkesworth, *Account,* pp. 548–49.

14. Ibid., p. 554; Cook, *Journals,* p. 348.

15. Bowen, *Great Barrier Reef,* p. 42; Banks, *Journal,* vol. 2, p. 113; Hawkesworth, *Account,* pp. 562–63.

16. Banks, *Journal,* vol. 2, p. 85.

17. Banks, quoted in Cook, *Journals,* fn. p. 352; Hawkesworth, *Account,* p. 561.

18. Cook, *Journals,* pp. 360–61; Hawkesworth, *Account,* p. 565; Banks, *Journal,* vol. 2, p. 87.

19. Banks, *Journal,* vol. 2, pp. 90, 94.

20. The suggestion of a turtle prohibition was made to me by the late Eric Deeral, MP, Guugu Yimithirr elder and clan custodian, when I visited Cooktown with the *Endeavour* replica in 2001, but I have been unable to verify it one way or another.

21. Cook, *Journals,* pp. 361–62; Hawkesworth, *Account,* p. 582.

22. Hawkesworth, *Account,* pp. 582–83.

23. This was also explained to me by Eric Deeral. It was later confirmed and elaborated on by another Guugu Yimithirr scholar, Alberta Hornsby, when we were making a documentary film version of Cook's environmental crisis at Endeavour River. See video interview with Alberta Hornsby, www.the-reef.com.au.

24. Banks, *Journal,* vol. 2, pp. 98–99.

25. Cook, *Journals,* p. 399. See also N. Thomas, *Cook: The Extraordinary Voyages of Captain James Cook,* New York, Walker and Co., 2003, p. 128.

26. Cook, *Journals,* pp. 396–97.

27. Ibid.

28. Ibid., p. 365; Hawkesworth, *Account,* p. 593.

29. Banks, *Journal,* vol. 2, pp. 101–103; Cook, *Journals,* p. 373.

30. Hawkesworth, *Account,* p. 601.

31. Ibid., p. 603.

32. Cook, *Journals,* pp. 378–79; Banks, *Journal,* vol. 2, p. 105.

33. Cook, *Journals,* p. 379; Banks, *Journal,* vol. 2, p. 106.

34. Banks, *Journal,* p. 107; Cook, *Journals,* pp. 379–80.

35. Banks, *Journal,* vol. 2, p. 108.

36. Ibid., p. 108; Cook, *Journals,* p. 380.

37. F. McLynn, *Captain Cook: Master of the Seas,* New Haven and London, Yale University Press, 2011, p. 153.

38. Cook, *Journals*, pp. 387–88.
39. Ibid., p. 390.
40. Ibid., p. 391.

2. BARRIER: MATTHEW FLINDERS'S DILEMMA

1. Flinders to George Bass, Reliance, Port Jackson, February 15, 1800, in M. Flinders, *Matthew Flinders: Personal Letters from an Extraordinary Life*, edited by P. Brunton, Sydney, Horden House in association with the State Library of New South Wales, 2002, p. 48; G. J. Barker-Benfield, "Sensibility," in I. McCalman, ed., *An Oxford Companion to the Romantic Age: British Culture, 1776–1832*, Oxford, Oxford University Press, 1999, pp. 102–14. See also J. Todd, *Sensibility: An Introduction*, London and New York, Methuen, 1986, pp. 3–31.
2. M. Fitzpatrick, "Enlightenment," and N. Thomas, "Exploration," in McCalman, *Oxford Companion*, pp. 299–311, 345–53.
3. Flinders to Sir Joseph Banks, Reliance, September 6, 1800, in *Personal Letters*, pp. 50–52.
4. Flinders to Ann, Port Jackson, July 20, 1802, in *Personal Letters*, pp. 84–86.
5. Flinders, *Personal Letters*, p. 25.
6. I. McCalman, *Darwin's Armada: How Four Voyagers to Australasia Won the Battle for Evolution and Changed the World*, Melbourne, Penguin, 2010, pp. 28–29; M. Flinders, *A Voyage to Terra Australis . . . 1801, 1802 and 1803 in His Majesty's Ship The Investigator . . .* 2 vols., plus atlas, London, G. and W. Nicol, 1814, vol. 1, pp. cxciii–cxciv. See also K. V. Smith, *Mari Nawi: Aboriginal Odysseys*, New South Wales, Rosenberg, 2010, pp. 102–17.
7. Smith, *Mari Nawi*, p. 102; Flinders, *Terra Australis*, vol. 2, p. 10.
8. Quoted in Smith, *Mari Nawi*, p. 105.
9. Flinders, *Terra Australis*, vol. 2, p. 10; Robert Brown's journal indicated that Bungaree could understand a few words of their language, R. Brown, *Nature's Investigator: The Diary of Robert Brown in Australia, 1801–1805*, compiled by T. G. Vallance, D. T. Moore, and E. W. Groves, Canberra, Australian Biological Resources Study, 2001, p. 232.
10. Flinders, *Terra Australis*, vol. 2, p. 30. See also the lively account of the incident by the gardener, Peter Good, *The Journal of Peter Good, Gardener on Matthew Flinders Voyage to Terra Australis 1801–1803*, ed. P. I. Edwards, London, British Museum, 1981, p. 86.
11. Flinders, *Terra Australis*, vol. 2, pp. 7, 10–11.
12. Ibid., pp. 238–39.
13. M. Estensen, *The Life of Matthew Flinders*, Sydney, Allen and Unwin, 2002, p. 239.
14. G. C. Ingleton, *Charting a Continent*, Sydney, Angus and Robertson, 1944, pp. 75–77.
15. Flinders, *Terra Australis*, vol. 2, pp. 7, 12, 13, 19.
16. Ibid., pp. 28–29, 36–39, 58.
17. Ibid., p. 43.
18. Ibid., pp. 67–70. The grueling circumstances of their surveys and scientific explorations of Broad Sound are best conveyed in Brown's diary, *Nature's Investigator*, pp. 251–82.
19. Flinders, *Terra Australis*, vol. 2, pp. 91, 84–85.
20. Ibid., p. 87. Brown, both more prosaic and grumpy, was simply annoyed that Flinders stayed too briefly on the reef for them to explore the different species of coral in more detail: *Nature's Investigator*, pp. 283–84.
21. Flinders, *Terra Australis*, vol. 2, p. 91.
22. Ibid., p. 92.
23. Ibid., pp. 109–111.
24. Ibid., p. 114.
25. Ibid., pp. 117–23.
26. Estensen, *Life of Matthew Flinders*, p. 45; Flinders, *Terra Australis*, vol. 2, p. 123.
27. Flinders, *Terra Australis*, pp. 141–43.

28. Ibid., p. 143.
29. J. D. Mack, *Matthew Flinders, 1774–1814*, Edinburgh and London, Thomas Nelson, 1966, pp. 145–46.
30. Quoted in Mack, *Flinders*, p. 146.
31. Estensen, *Life of Matthew Flinders*, p. 276.
32. For Brown's achievements on the voyage, see Bowen, *Great Barrier Reef*, pp. 65–73.
33. M. Flinders in *Philosophical Transactions* (Royal Soc.) 96 (1806), p. 252; entry in OED: "barrier reef," OED Online, September 2012, Oxford University Press, www.oed.com .ezproxy2.library.usyd.edu.au/view/Entry/15765?redirectedFrom=barrier+reef&, ac cessed October 17, 2012.
34. Flinders, *Terra Australis*, vol. 2, p. 102.
35. Ibid., p. 103.
36. Ibid., pp. 103–104.
37. Ibid., p. 115.
38. Ibid., pp. 115–16.
39. Ibid., p. 116.
40. McCalman, *Darwin's Armada*, p. 78.

3. CAGE: ELIZA FRASER'S HACK WRITER

1. The letter is reproduced in J. Curtis, *SHIPWRECK of the STIRLING CASTLE, . . . the Dreadful Sufferings of the Crew, . . . THE CRUEL MURDER OF CAPTAIN FRASER BY THE SAVAGES [and] . . . the Horrible Barbarity of the Cannibals Inflicted upon THE CAPTAIN'S WIDOW, Whose Unparalleled Sufferings Are Stated by Herself, and Corroborated by the Other Survivors*, London, George Virtue, 1838, pp. 206–207.
2. Quoted in M. Alexander, *Mrs. Fraser on the Fatal Shore*, London, Sphere Books, 1976, p. 130.
3. *The Times*, August 19, 1837, p. 6.
4. See L. T. Werkmeister, *A Newspaper History of England, 1792–1793*, Lincoln, University of Nebraska Press, 1967; L. T. Werkmeister, *The London Daily Press, 1772–1792*, Lincoln, University of Nebraska Press, 1963.
5. K. Schaffer, *In the Wake of First Contact: The Eliza Fraser Stories*, Cambridge, Cambridge University Press, 1995, pp. 68–69, 112–14. See also her incisive analysis, "Captivity Narratives and the Idea of Nation," in *Captured Lives: Australian Captivity Narratives*, eds., K. Darian Smith, R. Poignant, and K. Schaffer, London, Robert Menzies Centre for Australian Studies, 1993, pp. 1–13.
6. K. Wilson, *Island Race: Englishness, Empire and Gender in the 18th Century*, London, Routledge, 2003, ch. 5, passim. For something of the seamy milieu of nineteenth-century newsmen and publishers, see I. McCalman, *Radical Underworld: Prophets, Revolutionaries and Pornographers in London, 1795–1840*, Cambridge, Cambridge University Press, 1988, pp. 164–66.
7. Michael Alexander, who mentions James Curtis in passing, assumes that he is the same man as John Curtis, but cites no evidence, *Mrs. Fraser on the Fatal Shore*, p. 115. However, Stephen D. Behrendt's major Liverpool database of the period shows how commonly individuals switched freely between calling themselves James and John. See S. D. Behrendt et al., "Designing a Multi-Source Relational Database: 'Liverpool as Trading Port, 1700–1850,'" *International Journal of Maritime History*, vol. 24, no. 1 (June 2012), pp. 265–300. See also the description of James Curtis, in J. Grant, *The Great Metropolis*, 2 vols., London, Saunders and Otley, 1837, vol. 2, pp. 199–212; P. Collins, "The Molecatcher's Daughter," *The Independent*, November 26, 2006; L. James, *Print and the People, 1819–1851*, London, Viking, 1976, pp. 250–51; on Maria Marten, often also known as "The Red Barn Murder," see, R. D. Altick, *Victorian Studies in Scarlet*, New York and London, Norton, 1970, pp. 28–30, 93–96.

8. On Kelly and George Virtue, see J. C. Reid, *Bucks and Bruisers: Pierce Egan and Regency England*, London, Routledge, 1971, pp. 8, 73.

9. Curtis, *Shipwreck*, "Introduction," p. iv.

10. Ibid., pp. ii–v; Schaffer, *In the Wake*, pp. 49–60.

11. Schaffer, *In the Wake*, pp. 36–37, 45.

12. Curtis, *Shipwreck*, pp. 20–21.

13. Ibid., pp. 72–73.

14. Ibid., p. 147; *Colonial Times* (Hobart), January 16, 1838, pp. 5–6.

15. Curtis, *Shipwreck*, pp. 78–79, 155–56.

16. Curtis, *Shipwreck*, p. 193; *The Times*, August 19, 1837, p. 6.

17. Curtis, *Shipwreck*, pp. 41, 107, 115, 157–58. At this point he also inserted a footnote containing an assertion that Baxter and Mrs. Fraser had seen their captors, "after feeding on human food, carefully clean and preserve the bones of the victim."

18. Ibid., pp. 128–29.

19. Ibid., pp. 156–58.

20. Ibid., pp. 179–81.

21. Ibid., p. 186.

22. Ibid., p. 199; *Sydney Gazette*, January 25, 1838, p. 27, February 1, 1838, p. 2.

23. Curtis, *Shipwreck*, p. 209.

24. Ibid., pp. 204, 216–19.

25. Ibid., p. 215.

26. Ibid., fn. 219.

27. Darge's testimony is given in full in Alexander, *Mrs. Fraser*, pp. 119–25.

28. Curtis, *Shipwreck*, p. 221.

29. Ibid., pp. 222–23, 239, 241.

30. Ibid., p. 241. A detailed and accurate account of the Charles Eaton story has been published by V. Peek, "Voyage of the Charles Eaton/Charles Eaton Shipwreck," http://veronicapeek.com, accessed August 30, 2012.

31. J. Goodman, *The Rattlesnake: A Voyage of Discovery to the Coral Sea*, London, Faber, 2005, p. 8; on the influence of Barrow, see F. Fleming, *Barrow's Boys*, London, Granta, 1998; C. Lloyd, *Mr. Barrow of the Admiralty: A Life of Sir John Barrow*, London, Collins, 1970.

32. C. M. Lewis, *A Voyage to the Torres Strait in Search of the Survivors of the Charles Eaton, Which Was Wrecked upon the Barrier Reefs, in HM Colonial Schooner Isabella, C. M. Lewis Commander*, arranged by P. P. King, Sydney, E. H. Statham, 1837; *Sydney Times*, November 19, 1836; see also letter from J. W. Worthington, *The Times*, August 26, 1837, p. 5. Worthington, a friend of Bayley, also complained about Mayor Kelly having said that the Eliza Fraser case was one "of frightful novelty" when it was not. A brief account of the Mansion House interview with Ireland is in *The Times*, August 31, 1837, p. 6. For Ireland's rather incoherent memoir, see J. Ireland, *The Shipwrecked Orphans: A True Narrative of Four Years' Sufferings*, ed. Thomas Teller, New Haven, Babcock, 1845.

33. T. Wemyss, *Narrative of the Melancholy Shipwreck of the Charles Eaton*, Stockton, W. Robinson, 1837; V. Peek, "Charles Eaton Shipwreck," veronicapeek.com, accessed August 30, 2012.

34. T. B. Wilson, *Narrative of a Voyage Round the World: Comprehending an Account of the Wreck of the Ship, Governor Ready in the Torres Strait*, London, Gilbert and Piper, 1835; W. E. Brockett, *Narrative of a Voyage from Sydney to Torres Straits in Search of the Survivors of the Charles Eaton*, Sydney, Henry Bull, 1836.

35. Curtis, *Shipwreck*, p. 333.

36. Ibid., p. 256.

37. Ibid., pp. 324–25.

38. Wemyss, *Narrative*, pp. 34–35; Curtis, *Shipwreck*, pp. 336–46.

39. Curtis, *Shipwreck*, p. 308. Curtis commented on Duppa, "as far as that person (Duppa)

was concerned, the lad Ireland does not much complain"; Ireland, *Shipwrecked Orphans*, pp. 36–45; Lewis, *A Voyage to Torres Strait*, pp. 15–25.

40. Curtis, *Shipwreck*, pp. 302–303.

41. Ibid., p. 315.

42. Ibid., pp. 373–74.

43. Schaffer, *In the Wake*, chs. 6–10; R. Evans and J. Walker, " 'These Strangers, Where Are They Going?' Aboriginal–European Relations in the Fraser Island and Wide Bay Region, 1770–1905," *Occasional Papers in Anthropology*, no. 8 (March, 1977), pp. 42–45; J. Davidson, "Beyond the Fatal Shore: The Mythologization of Mrs. Fraser," *Meanjin*, 3 (1990), pp. 449–61.

4. BASTION: JOSEPH JUKES'S EPIPHANIES

1. J. Allen and P. Corris, eds., *The Journal of John Sweatman: A Nineteenth Century Surveying Voyage in North Australia and Torres Strait, St Lucia*, Queensland University Press, 1977, "Introduction," p. xvi.

2. J. B. Jukes, *Narrative of the Surveying Voyage of H.M.S. Fly . . . in the Torres Strait, New Guinea, and Other Islands of the Eastern Archipelago, during the years 1842–1846*, 2 vols., London, T. and W. Boone, 1847, "Appendix I," vol. 2, pp. 255–57.

3. Jukes to his cousin, Dr. Ingleby, July 12, 1848, in C. A. Browne, ed., *Letters and Extracts from the Addresses and Occasional Writings of J. Beete Jukes*, London, Chapman and Hall, 1871, p. 385.

4. Jukes, *Narrative of the Fly*, vol. 1, pp. 113–14.

5. H. S. Melville, *The Adventures of a Griffin on a Voyage of Discovery, Written by Himself*, London, Bell and Daldy, 1867, pp. 102–103.

6. Melville, *Adventures of a Griffin*, p. 105.

7. Jukes, *Narrative of the Fly*, vol. 1, pp. 77–83.

8. Melville, *Adventures of a Griffin*, pp. 5, 128.

9. Jukes, *Narrative of the Fly*, vol. 1, pp. 92, 112–14.

10. Quoted in Goodman, *The Rattlesnake*, p. 13.

11. Jukes, *Narrative of the Fly*, vol. 1, p. vi.

12. Ibid., pp. 8–9, 15–16.

13. Ibid., pp. 17–18.

14. Ibid., p. 316.

15. Ibid., p. 332.

16. Ibid., p. 347.

17. Jukes to Browne, November 26, 1843, Browne, *Letters and Addresses*, p. 215.

18. Melville, *Adventures of a Griffin*, pp. 179–82.

19. January 9–19, 1843, Browne, *Letters and Addresses*, pp. 173–74; Melville, *Adventures of a Griffin*, pp. 98–99.

20. Jukes, *Narrative of the Fly*, vol. 1, p. 118.

21. Browne, *Letters and Addresses*, pp. 8, 379–81, 401.

22. Ibid., p. 202.

23. Melville, *Adventures of a Griffin*, pp. 138–39.

24. Jukes, *Narrative of the Fly*, vol. 1, pp. 96–98; Melville, *Adventures of a Griffin*, p. 125.

25. R. Holmes, *Coleridge: Early Visions*, Harmondsworth, Penguin, 1990, p. 230.

26. Jukes, *Narrative of the Fly*, vol. 1, p. 122; E. Burke, *A Philosophical Enquiry into the Origin of Our Ideas of the Sublime and Beautiful*, London, R. and J. Dodsley, 1757.

27. Quoted in Melville, *Adventures of a Griffin*, pp. 126–27.

28. Melville, *Adventures of a Griffin*, p. 127; Jukes, *Narrative of the Fly*, vol. 1, pp. 122–24.

29. Jukes, *Narrative of the Fly*, pp. 48–49.

30. Ibid., pp. 145–53, 302–308.

31. Ibid., pp. 298–303.

32. Ibid., p. 297.

33. Jukes to Browne, August 27, 1843, Browne, *Letters and Addresses*, p. 198; Melville, *Adventures of a Griffin*, p. 190.

34. Melville, *Adventures of a Griffin*, pp. 197, 202–03; Jukes, *Narrative of the Fly*, vol. 1, pp. 155–80, 197–98.

35. Jukes, *Narrative of the Fly*, pp. 160–70, 209–10.

36. Ibid., p. 189.

37. Ibid., pp. 245–59.

38. Melville, *Adventures of a Griffin*, p. 205.

39. Browne, *Letters and Addresses*, pp. 247–248.

40. Ibid., pp. 249–50.

41. Jukes to Amelia Browne, October 2, 1845, Browne, *Letters and Addresses*, p. 263; see Melville's wistful account of his heroic colonial adventures, "Up Country in New South Wales," Melville, *Adventures of a Griffin*, pp. 225–48.

5. HEARTH: BARBARA THOMPSON, THE GHOST MAIDEN

1. D. R. Moore, *Islanders and Aborigines at Cape York: An Ethnographic Reconstruction Based on the 1848–1850 "Rattlesnake" Journals of O. W. Brierly and Information Obtained from Barbara Thompson*, Canberra, Australian Institute of Aboriginal Studies, 1979, pp. 8, 178. This splendid, meticulously accurate book consists of the full transcriptions of Brierly's interviews with Giom, and an insightful ethnographic survey of the Kaurareg.

2. Ibid., p. 186.

3. Ibid., p. 189.

4. Ibid., pp. 190–91.

5. Ibid., p. 160.

6. Ibid., p. 192.

7. Ibid., p. 194.

8. J. MacGillivray, *Narrative of the Voyage of HMS Rattlesnake . . . During the Years, 1846–1850*, 2 vols., London, Boone, 1852, vol. 1, p. 305.

9. Moore, *Islanders and Aborigines*, pp. 186–97.

10. Ibid., p. 76.

11. I am persuaded by the detailed detective work of R. J. Warren, *Wildflower: The Barbara Crawford Thompson Story*, Brisbane, R. J. Warren, rev. 3rd edition, 2012, pp. 35–36, that she was much younger than is generally thought, and probably shipwrecked at thirteen and rescued at the age of eighteen; MacGillivray, *Narrative of the Rattlesnake*, vol. 1, p. 305.

12. Moore, *Islanders and Aborigines*, p. 77.

13. Ibid., p. 77.

14. M. Bassett, *Behind the Picture: HMS Rattlesnake's Australia–New Guinea Cruise*, Melbourne, Oxford University Press, 1966, pp. 95–96; T. H. Huxley, *Diary of the Voyage of the Rattlesnake*, ed. J. Huxley, New York, Doubleday, 1936, p. 191.

15. Moore, *Islanders and Aborigines*, pp. 77–80, 224.

16. Ibid., p. 60.

17. Ibid., pp. 39–44.

18. Ibid., pp. 35–44, 54–70.

19. For a nuanced account of Brierly's relationships with Aborigines at Twofold Bay, see M. McKenna, *Looking for Blackfella's Point: An Australian History of Place*, Sydney, University of New South Wales Press, 2002, pp. 126–28; Goodman, *The Rattlesnake*, pp. 125–29.

20. Moore, *Islanders and Aborigines*, p. 35.

21. Ibid., pp. 46, 49.

22. Ibid., p. 64; Melville, *Adventures of a Griffin*, p. 182; O. W. Brierly, "Journal on HMS Rattlesnake. Torres Straits, New Guinea, . . . , 8 May 1849–8 Aug. 1849," Mitchell Library, State Library of New South Wales, MS A 507, mfm. MAV/FM4/2560.

23. Moore, *Islanders and Aborigines*, p. 197.

24. Huxley, *Diary of the Rattlesnake*, pp. 188–91; MacGillivray, *Narrative of the Rattlesnake*, vol. 1, p. 305.
25. Moore, *Islanders and Aborigines*, pp. 106–108, 146, 148–51, 155, 170, 176.
26. Ibid., pp. 150–53, 155–61, 173, 175, 182–5, 199–200, 203–204, 226.
27. Ibid., pp. 90, 121, 154, 164–66.
28. Ibid., pp. 91–92, 122, 116–19, 130, 166, 170.
29. Huxley, *Diary of the Rattlesnake*, p. 190; Moore, *Islanders and Aborigines*, pp. 80, 92, 162, 186.
30. Moore, *Islanders and Aborigines*, pp. 151, 213.
31. Ibid., pp. 121, 169.
32. Ibid., pp. 206–208.
33. Ibid., pp. 144–45, 177, 206–208.
34. MacGillivray, *Narrative of the Rattlesnake*, vol. 1, p. 302.
35. Ibid., pp. 306–307.
36. Moore, *Islanders and Aborigines*, "Introduction," pp. 8–9, 191.
37. Ibid., pp. 174, 190.
38. Ibid., pp. 163–67, 175, 179, 183.
39. Ibid., pp. 154–55, 170.
40. Ibid., p. 169; Warren, *Wildflower*, pp. 218–19, gives close attention to this issue, arguing that she also gave birth to a boy called Numa, but the evidence seems to me inconclusive.
41. Schaffer, *In the Wake*, esp. chs. 1–5.
42. For a wonderful study of the neglected naturalist's career and character, see Sophie Jensen, "On Such a Full Sea: John MacGillivray (1821–1867)," ANU, History PhD thesis, 2009.
43. MacGillivray, *Narrative of the Rattlesnake*, vol. 1, p. 305.
44. Warren, *Wildflower*, pp. 182–83. Ray Warren and Barbara's surviving relatives are to be congratulated for their considerable detective work in tracking these successive marriages; Moore, *Islanders and Aborigines*, pp. 12–13.

6. HEARTLANDS: THE LOST LIVES OF KARKYNJIB AND ANCO

1. L. de Rougemont, *The Adventures of Louis de Rougemont*, Project Gutenberg e-book, April 17, 2007, transcribed from the George Newnes edition, London, 1899.
2. B. G. Andrews, "Louis de Rougemont (1847–1921)," *Australian Dictionary of Biography*, Melbourne, Melbourne University Press, 1981, vol. 8; Rod Howard, *The Fabulist: The Incredible Story of Louis de Rougemont*, Sydney, Random House, 2006, pp. 224–36.
3. J. Morrill, *Seventeen Years Wandering Among the Aboriginals*, reprint of 1864 edn., Aboriginal Culture Series No. I, Virginia, NT, D. Welch, 2006, pp. 1–10.
4. Ibid., p. 11.
5. Ibid., pp. 7–21.
6. Ibid., p. 22; see also James Morrill, *Sketch of a Residence Among the Aboriginals of North Queensland*, Brisbane, Courier Office, 1863, pp. 8–10.
7. [Brisbane] *Courier*, March 11, 1863, p. 3.
8. S. Anderson, *Pelletier: The Forgotten Castaway of Cape York*, commentary on and translation of original book by Constant Merland, *Dix-sept ans chez les sauvages: Les aventures de Narcisse Pelletier* (1876), Melbourne, Melbourne Books, 2009, p. 327.
9. Merland, *Dix-sept*, in Anderson, *Pelletier*, pp. 34–45, 327.
10. Anderson, *Pelletier*, pp. 75–77.
11. Merland, *Dix-sept*, pp. 136–52.
12. Ibid., p. 150.
13. Ibid., pp. 153–54.
14. A. Chase, "Pama Malngkana," in Anderson, *Pelletier*, p. 108; Merland, *Dix-sept*, pp. 155–56.
15. H. Brayshaw, *Well Beaten Paths: Aborigines of the Herbert-Burdekin District of North*

Queensland: An Ethnographic and Archaeological Study, Townsville, James Cook University, 1990, pp. 12–13.

16. Morrill, *Seventeen Years*, p. 54.
17. Ibid., pp. 55–56.
18. Ibid., p. 70; [Brisbane] *Courier*, March 18, 1863, p. 3.
19. Chase, "Pama Malgnkana," pp. 91–94.
20. Ibid., pp. 92–95.
21. Merland, *Dix-sept*, pp. 238–72; on the subject of firestick farming, see Bill Gammage's important book, *The Biggest Estate on Earth: How Aborigines Made Australia*, Sydney, Allen and Unwin, 2011, passim.
22. Anderson, *Pelletier*, pp. 48, 57; Merland, *Dix-sept*, pp. 178, 266–72.
23. Merland, *Dix-sept*, p. 257.
24. Ibid., p. 181.
25. J. Farnfield, *Frontiersman: A Biography of George Elphinstone Dalrymple*, Melbourne, Oxford University Press, 1968, esp. pp. 58–62; G. C. Bolton, *A Thousand Miles Away: A History of North Queensland to 1920*, Canberra, Jacaranda Press, 1970, p. 27.
26. E. Dortins, "The Lives of Stories: Making Histories of Aboriginal-Settler Friendship," PhD thesis, University of Sydney, 2012, p. 14. I am grateful to Emma Dortins for allowing me to read and quote from this strikingly original and nuanced thesis; N. Loos, *Invasion and Resistance: Aboriginal-European Relations on the North Queensland Frontier, 1861–1897*, Canberra, Australian National University, 1982, pp. 28–42; B. Breslin, "Exterminate with Pride," *Aboriginal–European Relations in the Townsville–Bowen Region*, 1843–1869, Townsville, James Cook University, 1992, passim.
27. For a fine account of his experiences, see H. Reynolds, *The Other Side of the Frontier*, Townsville, James Cook University, 1981, pp. 14–16; Morrill, *Seventeen Years*, p. 42.
28. Morrill, *Seventeen Years*, pp. 42–43, 50; J. H. Peake, *A History of the Burdekin*, Ayr, Shire of Ayr Council, 1951, p. 4.
29. Morrill, *Seventeen Years*, p. 50.
30. Loos, *Invasion and Resistance*, pp. 118–22.
31. Anderson, *Pelletier*, pp. 42–43; S. Jensen, " 'On Such a Full Sea': John MacGillivray (1821–1867)," unpublished PhD thesis, ANU, 2009, pp. 6–11.
32. Anderson, *Pelletier*, p. 43. I am deeply grateful to Stephanie Anderson for finding and giving me access to the testimony of the *John Bell*'s captain, in her "Addendum," which has subsequently been published in the second revised paperback edition of her marvelous book and also in the electronic version. See Stephanie Anderson, *Pelletier: The Forgotten Castaway of Cape York*, 2nd revised edition, Melbourne, Melbourne Books, 2013, pp. 318–21.
33. *The Times*, July 21, 1875, quoted in Anderson, *Pelletier*, p. 45.
34. Merland, *Dix–sept*, p. 275; Anderson, *Pelletier*, p. 62.
35. Dortins, "Lives of Stories," pp. 25–28. For the broader context of Morrill's difficult circumstances, see H. Reynolds, "Aboriginal Resistance in Queensland," *Australian Journal of Politics and History*, vol. 22 (April, 1976), pp. 214–26.
36. See the sophisticated analysis of his ambivalence in Dortins, "Lives of Stories," pp. 14–15, 18.
37. Morrill, *Seventeen Years*, pp. 50–51.
38. Ibid., p. 71; Farnfield, *Frontiersman*, p. 69.
39. *Courier*, March 20, 1863, p. 3 and June 19, 1863, p. 2.
40. Farnfield, *Frontiersman*, p. 59; Morrill, *Seventeen Years*, pp. 70–71.
41. *Port Denison Times*, November 1, 1865.
42. S. Anderson, *Pelletier*, "The Two Lives of Narcisse Pelletier," p. 71, reports that this is modern oral tradition and folklore in Saint-Gilles.
43. Anderson, *Pelletier*, 2nd edition, "Addendum," extract from Louis de Kerjean, "Chroniques," *Revue de Bretagne et de Vendée*, 1876, pp. 324–25.

44. S. Anderson, "Three Living Australians and the Société d'Anthropologie de Paris, 1885," in *Foreign Bodies: Oceania and the Science of Race 1750–1840*, eds. B. Douglas and C. Ballard, Canberra, Australian National University Press, 2008, epress.anu.edu.au, accessed August 6, 2012, pp. 240–43.

45. Merland, *Dix-sept*, pp. 178, 186–91, 217–18.

46. Anderson, *Pelletier*, p. 46; Dortins, "Lives of Stories," pp. 33–34.

47. Anderson, *Pelletier*, p. 50; Morrill, *Seventeen Years*, p. 56. He noted that "human flesh cannot be considered a part of their food, although they sometimes eat it." Merland, *Dix-sept*, pp. 307–308.

48. Merland, *Dix-sept*, in *Pelletier*, pp. 167, 193, 199, 231–35, 273.

49. Howard, *The Fabulist*, pp. vii, 95, 251–52.

7. REFUGE: WILLIAM KENT ESCAPES HIS PAST

1. B. Taylor, *Cruelly Murdered: Constance Kent and the Killing at Road Hill House*, London, Souvenir, 1979, pp. 54–56.

2. C. Dickens, *The Mystery of Edwin Drood* (1870), London, Folio Society, 1982, p. 51.

3. These details were suggested in a letter, known as "The Sydney Document," sent anonymously by Constance Kent from Australia to the publisher Geoffrey Bles in 1929, and which, though now lost, is reproduced as appendix II, in Taylor, *Cruelly Murdered*, pp. 373–74. For a discussion of the syphilis accusation, see, K. Summerscale, *The Suspicions of Mr. Whicher, or The Murder at Road Hill House*, London, Bloomsbury, 2009, pp. 296–98.

4. Summerscale, *Suspicions*, pp. 13–29; Taylor, *Cruelly Murdered*, pp. 82–137.

5. N. Kyle, *A Greater Guilt: Constance Emilie Kent and the Road Murder*, Brisbane, Boolarong Press, 2009, pp. 150–53. I am grateful to Noeline Kyle for generously sharing her extraordinary knowledge of the case with me and for providing me with a manuscript copy of Constance Kent's "Sydney Document."

6. Kyle, *Greater Guilt*, pp. 225–29. She points out that most of the earlier analysts also believed that William was in some way involved, p. 225; Summerscale, *Suspicions*, pp. 299–302.

7. Kyle, *Greater Guilt*, p. 229; Taylor, *Cruelly Murdered*, pp. 296–322.

8. Constance Kent, "The Sydney Document" (1929), p. 5. This is a typescript MS copy lent to me by Noeline Kyle. It was obtained from a researcher, Stewart Evans, who transcribed the original for Taylor, before it was later lost.

9. See A. Desmond, *Huxley: From Devil's Disciple to Evolution's High Priest*, Reading, Massachusetts, Addison-Wesley, 1997, passim; I. McCalman, *Darwin's Armada*, pp. 293–339.

10. W. Saville-Kent to Sir William Flower, Brisbane, January 27, 1891, "Sir William Flower, Semi-Official Papers," *NLA* M2843, D932.

11. A. J. Harrison, *Savant of the Australian Seas: William Saville-Kent, 1842–1908*, Hobart, Tasmanian Historical Research Association, 2nd edition, n.d. [1997?], chs. 1–3.

12. W. Saville Kent, *A Manual of the Infusoria: Including a Description of All Known Flagellate, Ciliate and Tentaculiferous Protozoa*, 3 vols., London, D. Bogue, 1880–81; Desmond, *Huxley*, p. 533; Harrison, *Savant*, ch. 3b.

13. Harrison, *Savant*, ch. 3b; Kyle, *Greater Guilt*, pp. 29–31.

14. See the detailed research and analysis by Kyle, *Greater Guilt*, chs. 8–9, pp. 129–64.

15. See the excellent analysis of Saville-Kent in Tasmania, in Harrison, *Savant*, ch. 4; W. Saville-Kent, *The Great Barrier Reef of Australia: Its Products and Potentialities* (1893), facsimile edition, Melbourne, Currey, O'Neill, 1972, pp. 258–59.

16. Kyle, *Greater Guilt*, p. 150.

17. Taylor, *Cruelly Murdered*, pp. 143–46.

18. R. Brown, "Glimpses into the 19th Century Broadside Ballad Trade No 15: Constance Kent and the Road Murder," www.mustrad.org.uk/articles/bbals_15.htm, accessed May 20, 2013; see Noeline Kyle's discussion: *Greater Guilt*, pp. 153–57.

19. W. Saville-Kent, "Preliminary Observations on a Natural History Collection Made on the Surveying Cruise of HMS Myrmidon," *Papers and Proceedings of the Royal Society of Queensland*, vol. 6 (1889), p. 219.

20. Saville-Kent, *Great Barrier Reef*, p. viii.

21. Ibid., pp. 121–23; Saville-Kent, "Preliminary Observations," pp. 225–31.

22. Harrison, *Savant*, ch. 5.

23. R. Ganter, *The Pearl Shellers of Torres Strait: Resource Use and Decline, 1860s–1960s*, Melbourne, Melbourne University Press, 1994, esp. ch. 5, pp. 151–72; J.P.S. Bach, *The Pearling Industry of Australia*, Canberra, Commonwealth of Australia, 1955, pp. 42–60.

24. Harrison, *Savant*, ch. 6, p. 1 and also fn. 15. There is a photograph of the house in W. Saville-Kent, *The Naturalist in Australia*, London, Chapman and Hall, 1897, p. 39.

25. Saville-Kent, "Preliminary Observations," p. 220; see also "Works of Paul Foelsche (1813–1914)," artgallery.nsw.gov.au/collection/artist/foelsche-paul, accessed May 20, 2013.

26. Saville-Kent, *Naturalist in Australia*, pp. 39–51.

27. Harrison, *Savant*, ch. 6, p. 4, fn. 15; Kyle, *Greater Guilt*, pp. 157–59. Constance, under the name of Ruth Emilie Kaye, actually lived to receive a letter of congratulation from the Queen on her hundredth birthday.

28. Harrison, *Savant*, ch. 6, p. 3; Bach, *Pearling Industry*, pp. 54–55.

29. Saville-Kent, *Great Barrier Reef*, pp. 214–19.

30. Ganter, *Pearl Shellers*, pp. 168–71.

31. Saville-Kent, *Great Barrier Reef*, pp. 343–78.

32. Ibid., pp. 279–334.

33. Ibid., pp. 53, 334; Saville-Kent to Flower, January 27, 1891, "*Sir William Flower Papers*," D 932.

34. Saville-Kent, *Great Barrier Reef*, pp. 98–100.

35. Ibid., pp. 13–14, 19, 21, 37, 156, 159.

36. Ibid., pp. 139–40, 180.

37. Ibid., pp. 139–40.

38. A. Hingston Quiggin, *Haddon: The Head Hunter*, Cambridge, Cambridge University Press, 1942, p. 80.

39. Saville-Kent, *Great Barrier Reef*, pp. 32–33, 144–45.

40. Ibid., p. viii.

41. Ibid., p. 40.

42. Ibid., p. 124. For an excellent analysis of the multiracial and cultural makeup of Thursday Island at this time, see H. Reynolds, *North of Capricorn: The Untold Story of Australia's North*, Sydney, Allen and Unwin, 2003, pp. 85–103.

43. Saville-Kent, *Great Barrier Reef*, pp. 48, 59. On Frank Jardine, see J. Single, *The Torres Strait: People and History*, St. Lucia, University of Queensland Press, 1989, pp. 44–47, 55–56; C. Lack, "Jardine, Francis Lascelles (1841–1919)," *Australian Dictionary of Biography*, Australia National University, abd.anu.edu.au/biography/jardine-francis-lascelles-3924, accessed July 16, 2012.

44. *Argus*, August 23, 1896, p. 6; Harrison, *Savant*, ch. 6, p. 6; C. and M. Morton-Evans, *The Remarkable Life of Ellis Rowan*, Sydney, Simon and Schuster, 2008, pp. 11, 16. None of these authors hint at an affair; my suspicions are based on William's subsequent behavior, see below; V. Rae-Ellis, *Louisa Anne Meredith: A Tigress in Exile*, Hobart, Blubber Head Press, 1979, p. 210; J. B. Walker, *Prelude to Federation*, Hobart, Blubber Head Press, 1976, p. 115.

45. Love, *Reefscape*, 2001, pp. 97–104; Saville-Kent, *Great Barrier Reef*, pp. 25–26, 46, 49–50.

46. Saville-Kent, *Great Barrier Reef*, pp. 26–27.

47. *West Australian*, September 1, 1893, p. 7; "Extracts from Opinions of the Press," afterpiece in Saville-Kent, *Naturalist in Australia*.

48. Extracts from *The Times* and *Saturday Review*, Saville-Kent, *Naturalist in Australia*, p. 303. *Courier*, November 20, 1893, p. 6; *Argus*, July 29, 1893, p. 14.

49. "Extracts," Saville-Kent, *Naturalist in Australia*, ff. p. 302; *West Australian*, September 1, 1893.

50. Saville-Kent, *Naturalist in Australia*, p. 288.

51. Harrison, *Savant*, ch. 8 argues persuasively that William may indeed have preempted the Japanese in his quest.

8. PARADISE: TED BANFIELD'S ISLAND RETREAT

1. E. J. Banfield, *The Confessions of a Beachcomber* (1908), Rowville, Five Mile Press, 2006.

2. Banfield, "Dunk Island: Its General Characteristics," *Queensland Geographical Journal*, vol. 23 (1907–08), pp. 52–54.

3. Banfield, *Confessions*, pp. 21–22.

4. M. Noonan, *A Different Drummer: The Story of E. J. Banfield, Beachcomber of Dunk Island*, St. Lucia, University of Queensland Press, 1986, pp. 103–104.

5. Banfield, *Confessions*, p. 23.

6. E. J. Banfield, *My Tropic Isle*, London and Leipsic, T. Fisher Unwin, 1912.

7. Ibid., pp. 16–27; H. D. Thoreau, *Walden; or, Life in the Woods* (1854), New York, Dover, 1995.

8. Banfield, *My Tropic Isle*, p. 23.

9. Ibid., p. 51.

10. Banfield, *Confessions*, pp. 27–28; *My Tropic Isle*, p. 47.

11. Banfield, *My Tropic Isle*, pp. 29–30.

12. Banfield, *Confessions*, p. 299.

13. J. W. Banfield, "Reminiscences of an Incident at Dunolly Gold Rush and of Coming to Australia in 1852," Papers of Jabez Banfield, National Library of Australia, MS 1723, pp. 2–6.

14. Noonan, *Different Drummer*, pp. 28–30; R. F. Teichgraeber, *Sublime Thoughts/Penny Wisdom: Situating Emerson and Thoreau in the American Market*, Baltimore, Johns Hopkins, 1995, pp. 45–74.

15. For details of Ted's Townsville newspaper career, see J. Manion, *Paper Power in North Queensland: A History of Journalism in Townsville and Charters Towers*, Townsville, North Queensland Newspaper Company Ltd., 1982, pp. 177–79.

16. In return for the funds for his trip to England, Ted wrote a series of lively articles promoting the pleasures, safety, and convenience of Philp's British–India Steam Navigation Line, which used the swifter but riskier Torres Strait–Barrier Reef route. These were subsequently gathered into a much-feted local pamphlet, E. J. Banfield, *The Torres Strait Route: From Queensland to England*, Townsville, Thankful Willmett, 1885; E. J. to J. W. Banfield, November 5, 1890, Papers of Jabez Banfield, MS 1860, folder 4, Banfield Corr. 1–10.

17. See his retrospective letter to Harry Banfield, ibid., November 1, 1897.

18. Ted to Harry Banfield, ibid., Nov. 1, 1897; Noonan, *Different Drummer*, pp. 102–103.

19. Edmund Banfield's Diary 1898, Edmund Banfield's Collection, John Oxley Library, State Library of Queensland, Brisbane, Box 16031, pp. 16, 25, 33–36, 39, 41–42, 46–48, 52, 70; Diary 1899, pp. 5, 22, 26–27, 30–32, 71.

20. Ibid., p. 77; Noonan, *Different Drummer*, pp. 119–20.

21. See Diaries, 1898–1901, passim.

22. Noonan, *Different Drummer*, pp. 120–21.

23. See below, Chapter Eight, passim; Banfield's correspondents included F. Manson Bailey, author of *The Queensland Flora*, and the zoologist C. W. de Viss of the Queensland Museum; Banfield "Dunk Island: Its General Characteristics," pp. 59–60.

24. Banfield, *Confessions*, pp. 260–75.

25. C. Barrett, *Koonwarra: A Naturalist's Adventures in Australia*, London, Oxford University Press, H. Milford, 1939, p. 172.

26. Ibid., p. 173; Banfield, *My Tropic Isle*, "Swifts and Eagles," pp. 200–04.

27. Banfield, *Confessions*, p. 233.

28. Banfield, *Tropic Days* (1917), e-books, The University of Adelaide, 2010, "Blacks as Fishermen," pp. 107–20.

29. Banfield, *Confessions*, p. 260.

30. Banfield, *My Tropic Isle*, "Dead Finish," p. 281.

31. Ibid., pp. 28–29.

32. Banfield, "Dunk Island: Its General Characteristics," pp. 60–61.

33. Noonan, *Different Drummer*, pp. 148–50.

34. Banfield, *Confessions*, p. 130.

35. Ibid., p. 130.

36. Ibid., pp. 124–25.

37. Ibid., p. 146.

38. See, E. J. Banfield to Eliza, September 1915 to September 1918, "Letters, 1905–23," Papers of Jabez Banfield, MS 7105.

39. E. J. Banfield, *Last Leaves from Dunk Island*, South Yarra, Angus and Robertson, 1925, "The Tempest," pp. 1–2.

40. Ibid., p. 5.

41. Ibid., p. 11–13; Noonan, *Different Drummer*, p. 203.

42. Banfield, *Last Leaves*, p. 19.

43. The year before the publication of the *Confessions*, he also published the Queensland Government–funded, 103-page, illustrated *Within the Barrier: Tourists' Guide to the North Queensland Coast* (Brisbane, 1907).

44. Banfield, *Last Leaves*, pp. 182–88.

45. See A. H. Chisholm, *Birds and Green Places: A Book of Australian Nature Gossip*, London, J. M. Dent and Sons, 1929, esp. pp. 67–83; Noonan, *Different Drummer*, p. 228.

9. OBSESSION: THE QUEST TO PROVE THE ORIGINS OF THE REEF

1. A. G. Mayor, "Alexander Agassiz, 1835–1910," *Popular Science Monthly*, vol. 77 (November 1910), p. 434. Alfred Mayer later changed his surname to Mayor. Both spellings appear in his published works. I have used the later form throughout the text.

2. C. Darwin, *The Autobiography of Charles Darwin, 1809–82*, ed. N. Barlow, New York and London, Norton, 1958, p. 82.

3. Mayor, "Agassiz," p. 420.

4. D. Dobbs, *Reef Madness: Charles Darwin, Alex Agassiz and the Meaning of Coral*, New York, Pantheon Books, 2005, pp. 69–87.

5. Dobbs, *Reef Madness*, pp. 105–107.

6. December 22 and 24, 1873, Theodore Lyman Journals, Lyman Family Papers, 1785–1956, Massachusetts Historical Society, mfm, 1988, reel 23.

7. Dobbs, *Reef Madness*, p. 118; July 12, 1874, Theodore Lyman Journals, reel 23.

8. Mayor, "Agassiz," p. 425.

9. J. Murray, "Alexander Agassiz: His Life and Scientific Work," *Bulletin of the Museum of Comparative Zoology at Harvard College*, vol. 54, no. 3 (March, 1911), pp. 139–58.

10. Dobbs, *Reef Madness*, pp. 161–62; S. Jones, *Coral: A Pessimist in Paradise*, London, Little, Brown, 2007, pp. 33–34.

11. Murray, "Agassiz," *MCZ Bulletin*.

12. Dobbs, *Reef Madness*, pp. 167–69.

13. Agassiz to Haeckel, December 11, 1875, Agassiz Papers, Archives of the Ernst Mayr Library, Library of the Museum of Comparative Zoology, Harvard University, bAg. 10.10.19; A. Agassiz to Alex Braun, June 6, 1876, Alex Agassiz Letterbooks, Archives of the Ernst Mayr Library, box 45 A, slides 275–78.

14. Charles Darwin to Alexander Agassiz May 5, 1881, *Life and Letters of Charles Darwin*, ed. F. Darwin, 3 vols., London, Murray, 1887, vol. 3, pp. 183–84.

15. Dobbs, *Reef Madness*, pp. 193–95; T. H. Huxley, "Scientific and Pseudo-Scientific Realism" (1887) and "An Episcopal Trilogy" (1887), *Essays: Science and Christian Tradition*, New York, Appleton, 1896, pp. 59–159.

16. See McCalman, *Darwin's Armada*, pp. 78–81; Dobbs, *Reef Madness*, p. 199.

17. A. G. Mayor, "Expedition Journal—Around the World: Agassiz Expedition to the Great Barrier Reef," April 18, April 24–25, 1896 (no pagination), Alfred Goldsborough Mayor Papers, Syracuse University Library, Box 8.

18. A. Agassiz, "A Visit to the Great Barrier Reef of Australia in the Steamer 'Croydon,' during April and May, 1896," *Bulletin of the Museum of Comparative Zoology at Harvard College*, vol. 28, no. 4 (April 1898), pp. 103–104.

19. Agassiz, "Visit to the Great Barrier Reef," pp. 123, 143.

20. Ibid., pp. 132–39; see also, Mayor, "Journal," May 6, fos. 53–54.

21. Ibid., p. 143.

22. G. R. Agassiz, ed., *Letters and Recollections of Alexander Agassiz: With a Sketch of His Life and Work*, Boston and New York, Houghton Mifflin, 1913, p. 316; Mayor, "Journal," May 13–15, fos. 79–80.

23. December 3, 1897, Agassiz, *Letters and Recollections*, p. 329; Dobbs, *Reef Madness*, pp. 217–20.

24. Dobbs, *Reef Madness*, pp. 220–23; Agassiz to Murray, December 3, 1897, *Letters and Recollections*, pp. 328–29; Agassiz to Wolcot Gibbs, December 15, 1897, pp. 332–33.

25. Agassiz to Wolcot Gibbs, December 15, 1987, pp. 344–379, 394; see also, Mayor, "Agassiz," pp. 442–43.

26. Dobbs, *Reef Madness*, pp. 6–7.

27. T. C. Mendenhall, "Alfred Goldsborough Mayor," *Science*, new series, vol. 56, no. 1442 (Aug. 18, 1922), pp. 198–99.

28. For a description of Mayor, see C. B. Davenport, "Alfred Goldsborough Mayor," *Memoirs of the National Academy of Science*, vol. 21 (1926), p. 3; for Mayor's description of Agassiz as Bismarck, see Mayor, "Agassiz," pp. 439–40, 427.

29. R. S. Woodward, "Alfred Goldsborough Mayor," *Science*, vol. 56, no. 1438 (July 21, 1922) p. 68; Mayor quoted in Davenport's obituary article, "Alfred Goldsborough Mayor," p. 4.

30. Mayor, "Journal," April 25, 1896, Dunk Island (no pagination).

31. Ibid., May 3, 1896, Turtle Reef, fos. 35–37.

32. L. D. Stephens and D. R. Calder, *Seafaring Scientist: Alfred Goldsborough Mayor, Pioneer in Marine Biology*, Columbia, University of South Carolina Press, 2006, pp. 6, 12.

33. Ibid., pp. 12–16.

34. Mayor was much later, in 1915, to write a paper criticizing this idea, Stephens and Calder, *Seafaring Scientist*, p. 134. For a discussion of this theory from a modern expert, see S. Jones, *Coral*, p. 56.

35. Mayor, "Agassiz," p. 423; Stephens and Calder, *Seafaring Scientist*, pp. 41–42.

36. P. J. Bowler, *The Environmental Sciences*, London, Fontana, 1992, pp. 306–10; Mayor, "Agassiz," p. 427; Stephens and Calder, *Seafaring Scientist*, pp. 60–62.

37. Stephens and Calder, *Seafaring Scientist*, p. 15.

38. Davenport, "Mayor," p. 2.

39. Stephens and Calder, *Seafaring Scientist*, pp. 56–59.

40. Quoted in Davenport, "Mayor," p. 2.

41. Stephens and Calder, *Seafaring Scientist*, p. 62.

42. Mayor, "Agassiz," pp. 419–46.

43. A. G. Mayor, "An Expedition to the Coral Reefs of the Torres Straits," *Popular Science Monthly*, vol. 85 (September 1914), pp. 209–32.

44. A. G. Mayor, "Ecology of the Murray Island Coral Reef," *Papers from the Department of Marine Biology of the Carnegie Institution of Washington*, vol. 9, no. 213 (1918), Washington, DC, pp. 4–8.

45. Mayor, "Coral Reefs of the Torres Straits," pp. 214–15; Dobbs, *Reef Madness*, p. 251.

46. Mayor, "Ecology of the Murray Island Coral Reef," p. 29.

47. Mayor, "Coral Reefs of the Torres Straits," pp. 212–13.

48. Mayor, "Ecology of the Murray Island Coral Reef," p. 28; Stephens and Calder, *Seafaring Scientist*, p. 109.

49. Stephens and Calder, *Seafaring Scientist*, p. 144; Davenport, "Mayor," p. 10.

50. See S. Jones, *Coral*, pp. 36–39; Dobbs, *Reef Madness*, pp. 254–56.

10. SYMBIOSIS: CAMBRIDGE DONS ON A CORAL CAY

1. *Daily Express*, September 3, 1927; *Evening News* (London), September 3, 1927; *Sunday Express*, September 4, 1927; *Sunday Mail*, September 4, 1927; *Westminster Gazette*, September 13, 1927; *Daily Chronicle*, September 14, 1927.

2. C. M. Yonge, *A Year on The Great Barrier Reef: The Story of the Corals and the Greatest of Their Creations*, London and New York, Putnam, 1930, p. 24. He mentioned having seen depictions of the Barrier Reef through Saville-Kent's remarkable photographs.

3. *The Times*, April 2, 1927; *Westminster Independent*, September 11, 1927; *Westminster Gazette*, September 13, 1927.

4. M. J. Yonge, "Log of Outward Journey to Great Barrier Reef, 28 May 1928–4 July 1928," Sir Maurice Yonge (1899–1986): Australian and New Zealand papers, National Library of Australia, mfm M2844–2846, p. 50 (my pagination).

5. *Sydney Morning Herald*, July 5, 1928, p. 10.

6. For Mel Ward, see Mel Ward Papers, Australian Museum, AMS 230, Box 12, folder 96, *Sunday Times*, September 18, 1927, "Theatrical Star as Scientist"; and *People*, January 17, 1951, "He Knows How to Bake a Snake," pp. 26–29; M. Murray and J. Roach, "Whitley, Gilbert Percy (1903–1975)," *Australian Dictionary of Biography*, www.adb.anu.edu.au /biography/whitley-gilbert-percy-12022/text21563, accessed November 15, 2012; on Frank McNeill, information was provided to the author by the reference staff of the Australian Museum.

7. *Sun*, July 6, 1928.

8. H. C. Richards, "Problems of the Great Barrier Reef," *Queensland Geographical Journal*, vol. 37 (1922), pp. 42–54; D. Hill, "The Great Barrier Reef Committee, 1922–82: The First Thirty Years," *Historical Records of Australian Science*, vol. 6, no. 1, pp. 1–4; Bowen, *Great Barrier Reef*, pp. 240–50.

9. Bowen, *Great Barrier Reef*, p. 254.

10. Ibid., p. 250.

11. S. Elliott Napier, *On the Barrier Reef* [1928], Sydney and London, Angus and Robertson, 1939 reprint, pp. 1–8. For his tribute to Charles Hedley, see fn. p. 195.

12. B. E. Brown, "The Legacy of Professor John Stanley Gardiner, FRS, to Reef Science," *Notes and Records of the Royal Society*, vol. 61 (May 22, 2007), pp. 207–16; quoted in Bowen, *Great Barrier Reef*, p. 244.

13. Gardiner's epistolary machinations can be seen in "Yonge Barrier Reef corres," Sir Maurice Yonge Papers, and they are also splendidly analyzed in Bowen, *Great Barrier Reef*, pp. 252–58; see also B. Morton, "Charles Maurice Yonge," *Biographical Memoirs of the Royal Society*, vol. 38 (1992), pp. 386–89.

14. Brown, "Legacy of Gardiner," pp. 207–17; G. Bidder to M. Yonge, July 8, 1927, "Yonge Barrier Reef corres," Sir Maurice Yonge Papers, E 49.

15. Charles Maurice Yonge Journal, "Great Barrier Reef Expedition, 1928–29," Sir Maurice Yonge Papers, fo. 9 and M. J. Yonge Journal, Sir Maurice Yonge Papers, B.4, fo. 2.

16. C. Barrett, "Science in Shirt Sleeves," *Adelaide Register*, August 14, 1928, p. 9. For example, he stressed the potential importance of the collecting work to be done by the Australian Museum scientists, and trumpeted the existing geological achievements of

Richards and his colleagues in having demonstrated the applicability of Darwin's theory of subsidence to the Reef's origins.

17. Yonge, *A Year on the Reef*, p. 97.
18. C. Barrett, "Scientists in Wonderland," *Adelaide Register*, August 1, 1928, p. 9; "On a Coral Isle: Peering at Barrier Reef Secrets," *Sydney Morning Herald*, August 2, 1928, p. 10; "On a Coral Isle: Scientific Research," *Sydney Morning Herald*, August 8, 1928, p. 14; "Science in Shirt Sleeves," p. 9.
19. Barrett, "Scientists in Wonderland," p. 9; "Peering," p. 10.
20. Yonge, *A Year on the Reef*, pp. viii, 102.
21. Barrett, "Peering," p. 10.
22. Barrett, "Pastures of the Sea," *Burnie Advocate*, August 29, 1928, p. II.
23. Barrett, "Scientific Research," p. 14.
24. Barrett, "On a Coral Isle: Barrier Reef Theories," *Sydney Morning Herald*, August 20, 1928, p. 15.
25. Barrett, "On a Coral Isle: Barrier Reef Fish and Snakes," *Sydney Morning Herald*, August 25, 1928, p. 13; "Barrier Reef Theories," p. 15.
26. Charles Barrett also published a longer version of his articles, with substantial illustrations, "The Great Barrier Reef and Its Isles: The Wonder and Mystery of Australia's World-Famous Geographical Feature," *National Geographic Magazine*, vol. 58 (1930), pp. 354–84.
27. C. M. Yonge Journal, Sir Maurice Yonge Papers, Oct. 9, 1928, fo. 55; Yonge, *A Year on the Reef*, p. 36.
28. Yonge, *A Year on the Reef*, pp. 28–30; Bowen, *Great Barrier Reef*, p. 259.
29. Brown, "Legacy of Gardiner," pp. 207–17.
30. Yonge, *A Year on the Reef*, pp. 58–59, 109.
31. See the delightful discussion in Jones, *Coral*, pp. 95–102; Yonge, *A Year on the Reef*, pp. 109–10.
32. Yonge, *A Year on the Reef*, p. 111.
33. The work of Tom Goreau, Von Holt and Von Holt, and Muscatine and Cernchiari was especially influential in this revision: see, Brown, "Legacy of Gardiner," pp. 215–17; Bowen, *Great Barrier Reef*, p. 282.
34. Bowen, *Great Barrier Reef*, p. 282; Jones, *Coral*, pp. 104–105.
35. Jones, *Coral*, p. 105.
36. R. Love and N. Grasset, "Women Scientists and the Great Barrier Reef: A Historical Perspective," *Search*, vol. 15 (October–November, 1984), pp. 285–86; Sir F. Russell, "Sheina Macalister Marshall, 20 April 1896–7 April 1977," *Biographical Memoirs of Fellows of the Royal Society*, vol. 24 (1978), pp. 368–89.
37. C. M. Yonge Journal, Sir Maurice Yonge Papers, June 4, 1929, fo. 122; M. J. Yonge Journal, Sir Maurice Yonge Papers, July 10, 1828, pp. 57–60 (my pagination).
38. See, for example, *Sydney Morning Herald*, August 8, 1929, p. 10; August 30, 1929, p. 10; C. M. Yonge, Great Barrier Reef Expedition, Log of Trip to Torres Straits, April–May 1929, July 10, 1828, Sir Maurice Yonge Papers, fos. 1–28; Bowen, *Great Barrier Reef*, pp. 278–79.
39. Yonge, *A Year on the Reef*, p. 203.
40. Ibid., p. 146.
41. E. M. Embury, *The Great Barrier Reef*, Sydney, Shakespeare Head Press, 1933, p. 19; Anon., "The Embury Story," Embury scientific and holiday expeditions on the Great Barrier Reef, Mitchell Library, State Library of New South Wales, MSS PXA 642/185, pp. 1–2.
42. Napier, *On the Barrier Reef*, passim; E. F. Pollock, *Expeditions to Great Barrier Reef Islands Off the Queensland Coast*, Strathfield, New South Wales, E. F. Pollock, n.d.; Arch Embury to Marion Mahon, March 1981, Embury scientific and holiday expeditions,

187, p. 2; "Members of Lindeman Island Great Barrier Reef Expedition," Whitsunday Passage, Queensland, December 17, 1928–January 19, 1929, Newcastle, Embury scientific and holiday expeditions, 182; Anon., "The Embury Story," p. 3.

43. Bowen, *Great Barrier Reef*, p. 285.

44. Anon., "The Embury Story," p. 3; F. A. McNeill, Newscutting Books, Australian Museum, AN90/72, Book I, Island of Desire," July 1, 1932; *Sun*, April 24, 1932; *Brisbane Courier*, May 16, 1932; "Arch Helped Pioneer Tourism on the Barrier Reef," *Weekend Feature*, n.d.; "Rare Coral Looted," *Sun*, January 31, 1932; Book 2, "Sydney Scientists and Tourists," *Telegraph*, December 6, 1932; "Scientific Party's Plan," *Sun* December 1, 1932; *Brisbane Telegraph*, December 22, 1932; *Herald*, February 23, 1933; "The Barrier Reef Should Be Better Protected," *Sun*, February 21, 1933; Frank McNeill, "The Story of Coral," *Bank Notes*, October 1932; Embury, *Great Barrier Reef*, pp. 72, 77, 83.

45. F. A. McNeill, "Newscutting Books," Book 1, "Lionel G. Wigmore Describes, Mysteries of the Great Barrier Reef," *Sunday Sun and Guardian*.

II. WAR: A POET, A FORESTER, AND AN ARTIST JOIN FORCES

1. John Busst to Judith Wright, September 2, 1967, John Busst Collection, [JBC] James Cook University, JBC/Corr/13.

2. P. Clare, *The Struggle for the Great Barrier Reef*, London, Collins, 1971, pp. 88–92.

3. J. Wright, *The Coral Battleground*, Sydney, Angus and Robertson, 1996, p. 4.

4. Clare, *Struggle*, pp. 94–97.

5. J. Busst to L. W. Webb, October 9 [1967], JBC/Corr/13.

6. Clare, *Struggle*, pp. 97–98.

7. Ibid., pp. 99–104.

8. See N. Monkman, *From Queensland to the Barrier Reef*, New York, Doubleday and Co., 1958; and *Quest of the Curly Tailed Horses*, Sydney, Angus and Robertson, 1962; Clare, *Struggle*, pp. 55–74.

9. R. Carson, *The Sea*, London, MacGibbon and Kee, 1964, containing *The Sea Around Us*, *Under the Sea-Wind*, *The Edge of the Sea*; R. Carson, *Silent Spring*, Boston and New York, Houghton Mifflin, 1962.

10. "Queensland Littoral Society," *Wildlife*, no. 7 (June 1967).

11. Bowen, *Great Barrier Reef*, pp. 319–21.

12. Wright, *Coral Battlefield*, p. 24; J. Busst to the *Advocate*, n.d. (c. April 1967), JBC/PRS/8.

13. V. Brady, *South of My Days: A Biography of Judith Wright*, Sydney, Angus and Robertson, 1998, pp. 21–27, 189–91; P. Clarke and M. McKinney, *With Love and Fury: Selected Letters of Judith Wright*, Canberra, National Library of Australia, 2006, p. xi.

14. Brady, *South of My Days*, pp. 112–13, 191.

15. Wildlife Preservation Society of Queensland Historical Papers. *Catalyst for Action: Formation of a Conservation Society*, Monograph 1, July 2006, www.wildlife.org.au, accessed November 8, 2012. For the influence of McKinney, Walker, and MacArthur, see Brady, *South of My Days*, pp. 11–27, 170–86; Emma Dortins, "The Lives of Stories: Making Histories of Aboriginal-Settler Friendship," pp. 204–50.

16. Wright, *Coral Battleground*, p. 1.

17. J. Wright, "The Builders," *Collected Poems, 1942–85*, Sydney, Angus and Robertson, 1994, p. 45.

18. J. Wright to Kathleen MacArthur, November 20, 1962; Clarke and McKinney, *With Love and Fury*, pp. 147–48; Wright, *Coral Battleground*, p. 3.

19. G. Tracey, "Len Webb: Pioneer in Ecology of the Rainforests of Australia," *Australian Science Magazine*, vol. 4 (1988), p. 66.

20. Ibid., pp. 66–67.

21. I. Frazer, "Conservation and Farming in Northern Queensland, 1861–1970," MA (research), James Cook University, pp. 163–64, eprints.jcu.au/78, accessed August 5, 2012.

22. Tracey, "Len Webb," pp. 66–70; Frazer, "Conservation and Farming," pp. 165–66; Wildlife Preservation Society of Queensland Historical Papers, *Heart and Mind: WPSQ Finding Directions in the 60s*, Monograph 2, May 2008, www.wildlife.org.au, accessed November 8, 2012; "John Geoffrey Tracey, The Field Botanist," in G. Borschmann, *The People's Forest: A Living History of the Australian Bush*, Blackheath, NSW, People's Forest Press, 1999, pp. 219–20.

23. Frazer, "Conservation and Farming," p. 163; see also Busst's retrospective reflections in J. Busst to Len Webb, July 22, 1968, JBC/Corr/6.

24. B. Roland, *The Eye of the Beholder*, Sydney, Hale and Iremonger, 1984, pp. 14, 30, 44, 172–73.

25. Clare, *Struggle*, p. 89; Roland, *Eye of the Beholder*, p. 244.

26. Frazer, "Conservation and Farming," p. 161; Len Webb to Ali and John Busst, November 8, 1966, JBC/Corr/3.

27. Frazer, "Conservation and Farming," pp. 152–53; Clare, *Struggle*, pp. 89–90.

28. Wright, *Coral Battlefield*, p. 188; Frazer, "Conservation and Farming," pp. 169–70; Barry Wain, "The Bingal Bay Bastard," *Nation* (May 1, 1971), p. 14.

29. John Busst to Len Webb, November 12 [1966/7?], JBC/Corr/3; Wright, *Coral Battleground*, pp. 13, 188–89. This point is also well made in Phoebe Ford's excellent History IV Hons thesis, "Consilience: Saving the Great Barrier Reef, 1962–1975," which I was privileged to supervise.

30. L. Webb, "Notes for an Address to the Students at Kendron Teachers College," *Wildlife*, no. 7 (June, 1967); WPSQ Historical Papers, *Heart and Mind*, p. 2.

31. J. Wright, *Because I Was Invited*, Melbourne, Oxford University Press, 1975, pp. 189–94.

32. Clare, *Struggle*, pp. 10–11; Wright, *Coral Battlefield*, p. 23.

33. See Bowen, *Great Barrier Reef*, pp. 318–21; W. J. Lines, *Patriots: Defending Australia's Natural Heritage*, St. Lucia, University of Queensland Press, 2006, pp. 146–47; Brady, *South of My Days*, pp. 242–44, 255–59.

34. Wright, *Coral Battleground*, pp. 17–18.

35. Clare, *Struggle*, pp. 114–15; Bowen, *Great Barrier Reef*, pp. 321–23.

36. J. Busst to Barry Wain, August 3, 1968, JBC/Corr/8; Wright, *Coral Battleground*, p. 34; Lines, *Patriots*, pp. 78–80; Wright, *Coral Battleground*, pp. 20, 65.

37. Tracey, "Len Webb," p. 68.

38. L. Webb and J. W. T. Williams, "Synecology—Cinderella Finds Her Coach," *New Scientist*, no. 59 (July 26, 1973), pp. 195–96.

39. See draft letter of J. Busst to President Lyndon Johnson, February 12, 1968, JBC/Corr/18; J. Busst to J. G. Gorton, March 17, 1969, JBC/Corr/9; November 18, 1969, JBC/Corr/12; March 18, 1971, JBC/Corr/12; J. Busst to Gough Whitlam, August 26, 1968, JBC/Corr/16; January 13, 1969, JBC/Corr/7; June 26, 1969, JBC/Corr/9.

40. Wright, *Collected Poems*, pp. 287–88.

41. Brady, *South of My Days*, p. 235.

42. Wright, "Education and Environmental Crisis"; *Because I Was Invited*, pp. 220–21.

43. Wright, *Coral Battlefield*, p. 103; Brady, *South of My Days*, p. 239.

44. Judith Wright, "The Quiet Crisis of Our Time," *Outlook*, no. 3 (June, 1969), pp. 3–5; Wright, *Coral Battlefield*, p. 186; Judith Wright, "The Role of Public Opinion in Conservation," in *The Last of Lands: Conservation in Australia*, eds. L. J. Webb, D. Whitelock, J. Le Gay Brereton, Milton, Queensland, Jacaranda, 1969, pp. 43–51; Wright, "Science, Value and Meaning," *Because I Was Invited*, pp. 196–202.

45. Wright, *Coral Battleground*, pp. 53–54.

46. Ibid., p. 105.

47. Bowen, *Great Barrier Reef*, pp. 334–35; Ford, "Consilience," pp. 80–82.

48. Bowen, *Great Barrier Reef*, pp. 331–78.

49. Wain, "Bingal Bay Bastard," p. 14.

50. Dortins, "Lives of Stories," pp. 204–14.
51. Wright to Professor J. R. Burton, University of New England, April 1, 1971, *With Love and Fury*, pp. 218–20; Len Webb, "The Plight of the Plant Ecologist," Draft Address to ANZAAS, Brisbane, May 1971, Len Webb Collection [LWC], Griffith University, Unpublished Papers and Reports, Box 158, folder 161, p. 7; see also Ford, "Consilience," pp. 87–94.

12. EXTINCTION: CHARLIE VERON, DARWIN OF THE CORAL

1. J. E. N. Veron, "Is the Great Barrier Reef on Death Row?," lecture at the Royal Society, London, July 6, 2009, www.royalsociety.org/events/2009/barrier-reef/, accessed October 10, 2012.
2. J. E. N. Veron, *A Reef in Time: The Great Barrier Reef From Beginning to End*, Cambridge, MA, Harvard University Press, 2008, p. vii.
3. "Charlie Veron's Story," unpublished autobiography, December 2011, pp. 12–13. I am deeply grateful to Charlie Veron for allowing me to read and quote from this compelling work, which was written for his children. It is a privilege and a responsibility. He cannot be held responsible, however, for my interpretation of his life.
4. See, "Tribute: Dr. Isobel Bennett, AO," *WISENET Journal*, no. 44 (July 1997), pp. 1–3, www.wisenet-australia.org/issue77/Isobel Bennett.htm, accessed May 1, 2013; "Biography— Dr Isobel Bennett AO," Australian Government: Great Barrier Marine Park Authority, www.gbrmpa.gov.au/resources-and-publications/spatial-data-information-services /reefinfo/isobel-bennett-reef/biography-dr-isobel-bennett-ao, accessed October 10, 2012; A. Nessy, "The Sea Has Many Voices: Profile of an Australian Woman Scientist," *Journal of Australian Studies*, vol. 38 (1992), pp. 41–50; Janet Browne, *Voyaging*, pp. 117–34.
5. "Charlie Veron's Story," pp. 20, 30–35.
6. Ibid., pp. 35–36.
7. Ibid., pp. 19–25.
8. Ibid., pp. 39–40; for Darwin, see *Autobiography of Charles Darwin*, p. 49; Browne, *Voyaging*, pp. 89–90, 145–46.
9. "Charlie Veron's Story," p. 49.
10. Ibid., pp. 51–52, 67.
11. Ibid., pp. 51–52, 56–59; on Wallace, see McCalman, *Darwin's Armada*, pp. 221–44.
12. "Charlie Veron's Story," pp. 59–60.
13. Ibid., p. 62.
14. Ibid., p. 63.
15. Ibid., p. 74.
16. Ibid., pp. 75–76.
17. Ibid., p. 78.
18. "1996 AMSA Silver Jubilee Awardee, Dr Charlie Veron," reprinted from *Golden Anniversary Issue of the Atoll Research Bulletin*, vol. 494 (2001), pp. 109–117, www.amsa.aen-au /awards/winners_silverjubilee/1996_veron_charlie.php.
19. "Charlie Veron's Story," pp. 80–82.
20. Ibid., pp. 80–84.
21. Ibid., pp. 84–85.
22. Ibid., p. 121.
23. Ibid., pp. 121–23.
24. J. E. N. Veron and G. Borschmann, *J. E. N. "Charlie" Veron Interview*, Townsville, February 28, 2005, and April 1–3, 2005, National Library of Australia, Oral History Project, Tape 3.
25. "Charlie Veron's Story," pp. 129–30; James Hamilton-Paterson has also written movingly about his similar ritual of taking night dives on reefs: *Seven-Tenths: The Sea and Its Thresholds*, New York, Europa Editions, 2009, pp. 137–45.

26. Browne, *Voyaging*, pp. 441–43.
27. Veron and Borschmann, *Interview*, NLA, Tape 3; "Charlie Veron's Story," p. 86.
28. Veron Interview, *1996 AMSA Silver Jubilee Awardee*, p. 4.
29. J. E. N. Veron, interviewed by Gregg Borschmann, "The Songlines Conversations: John (Charlie) Veron," *Big Ideas*, ABC Radio, August 6, 2006, www.abc.net.au/radionational /programs/bigideas/the-songlines-conversations-john-charlie-veron/3338280, accessed October 10, 2012; "Charlie Veron's Story," p. 106.
30. Veron interview, *1996 AMSA Silver Jubilee Awardee*, p. 4; "Charlie Veron's Story," pp. 116–17.
31. "Charlie Veron's Story," pp. 155–59.
32. Ibid., pp. 207–11, 214–15.
33. Veron interview, *1996 AMSA Silver Jubilee Awardee*, p. 6; Veron, ABC Interview, *Big Ideas*, p. 13; "Charlie Veron's Story," pp. 175–76.
34. "Charlie Veron's Story," p. 211.
35. Veron interview, *1996 AMSA Silver Jubilee Awardee*, p. 5.
36. Veron, ABC interview, *Big Ideas*, p. 12.
37. "Charlie Veron's Story," pp. 207–16; J. E. N. Veron, *Corals in Space and Time: The Biogeography and Evolution of the Scleractinia*, Sydney, University of New South Wales Press, 1995, see especially Part D, Evolution, chs. 12–13.
38. M. L. Arnold and N. D. Fogarty, "Reticulate Evolution and Marine Organisms: The Final Frontier?" *International Journal of Molecular Sciences*, vol. 10, no. 9, pp. 3836–3860.
39. "Charlie Veron's Story," pp. 11, 102.
40. Ibid., p. 136.
41. Ibid., p. 144.
42. Darwin, *Life and Letters*, vol. 1, pp. 132–34.
43. "Charlie Veron's Story," p. 146.
44. Veron, ABC Interview, *Big Ideas*, p. 16.
45. Veron interview, *1996 AMSA Silver Jubilee Awardee*, p. 8.
46. "Charlie Veron's Story," pp. 215–17.
47. Veron, ABC interview, *Big Ideas*, pp. 17–18. Given that the recent AIMS survey of the Barrier Reef contends that 42 percent of the present coral loss still comes from this voracious starfish, Charlie's fury was not misplaced. "The Great Barrier Reef has lost half of its coral in the last twenty-seven years," *AIMS News*, October 2, 2012, www.aims.gov.au, accessed October 3, 2012.
48. Veron, ABC interview, *Big Ideas*, pp. 5, 6.
49. Ibid., p. 6.
50. Veron, *A Reef in Time*, pp. 57–59.
51. The just released AIMS survey calculates that 10 percent of Barrier Reef coral loss can be attributed to climate change, but, like Charlie, it anticipates sharply increasing incidences of excess warming, *AIMS News*, October 2, 2012.
52. Veron, *A Reef in Time*, pp. 56–65, 200–11.
53. Ibid., pp. 89–112.
54. Ibid., pp. 214–16.
55. *AIMS News*, October 2, 2012.
56. Veron, *A Reef in Time*, pp. 212–20, 231.
57. Veron, ABC interview, *Big Ideas*, p. 3.
58. Veron, "Death Row."

EPILOGUE: A COUNTRY OF THE HEART

1. T. Horwitz, *Into the Blue: Boldly Going Where Captain Cook Has Gone Before*, Sydney, Allen and Unwin, 2002, pp. 170–74, 195–98.

2. Sadly, Eric Deeral died in Hope Vale, Queensland, on September 5, 2012, at the age of 80.

3. Banfield, *Last Leaves*, p. 11.

4. "Net Benefits," ABC Broadcast, July 4, 2010, www.abc.net.au/landline/content/2010 /s2944218.htm, accessed June 7, 2013.

BIBLIOGRAPHY

Newspapers
Adelaide Register
Argus
Burnie Advocate
Courier (Brisbane)
Daily Chronicle
Daily Express
Evening News (London)
The Independent
Port Denison Times
The Sun
Sunday Express
Sunday Mail
Sydney Gazette
Sydney Morning Herald
Sydney Times
The Times (London)
West Australian
Westminster Gazette
Westminster Independent

Unpublished Sources
Agassiz, A., Agassiz Papers, Archives of the Ernst Mayr Library, Library of the Museum of
  Comparative Zoology, Harvard University
———. Letterbooks, Archives of the Ernst Mayr Library, Library of the Museum of Com-
  parative Zoology, Harvard University
Banfield, E., Diaries, Edmund Banfield's Collection, John Oxley Library, State Library of
  Queensland, Brisbane, Box 16031

Banfield, J., Papers of Jabez Banfield, 1781–1897, National Library of Australia, MS 1860, MS 7105 and MS 1723

Brierly, O. W., "Journal on HMS Rattlesnake. Torres Straits, New Guinea, . . . , 8 May 1849–8 Aug. 1849," Mitchell Library, State Library of New South Wales, MS A 507, mfm MAV/FM4/2560

Busst, J., "John Busst Collection," [JBC] James Cook University

Dortins, E., "The Lives of Stories: Making Histories of Aboriginal-Settler Friendship," PhD thesis, University of Sydney, 2012

Embury, A. and M., Embury scientific and holiday expeditions on the Great Barrier Reef, Mitchell Library, State Library of New South Wales, MSS PXA 642/185

Flower, Sir William, Semi-Official Papers, NLA M2843, D932

Ford, P., "Consilience: Saving the Great Barrier Reef, 1962–1975," hons thesis, University of Sydney, 2012

Frazer, I., "Conservation and Farming in Northern Queensland, 1861–1970," MA (research), James Cook University, pp. 163–64. http://eprints.jcu.au/78

Jensen, S., "On Such a Full Sea: John MacGillivray (1821–67)," history PhD thesis, Australian National University, 2009

Kent, Constance, "The Sydney Document" (1929)

Lyman, T., Theodore Lyman Journals, Lyman Family Papers, 1785–1956, Massachusetts Historical Society, mfm, 1988, reel 23

Mayor, A. G., "Expedition Journal—Around the World: Agassiz Expedition to the Great Barrier Reef," April 18, April 24–25, 1896 (no pagination), Alfred Goldsborough Mayor Papers, Syracuse University Library, Box 8

McNeill, F. A., Newscutting Books, Australian Museum, AN90/72

Veron, J. E. N., "Charlie Veron's Story," unpublished autobiography, Dec. 2011

Veron, J. E. N. and G. Borschmann, J. E. N. "Charlie" Veron interview, Townsville, Feb. 28, 2005, and April 1–3, 2005, National Library of Australia, Oral History Project

Ward, M., Mel Ward Papers, Australian Museum, AMS 230, Box 12

Webb, L., "Len Webb Collection" [LWC]. Griffith University, Unpublished Papers and Reports

Yonge, C. M., Sir Maurice Yonge (1899–1986): Australian and New Zealand papers, National Library of Australia, mfm M2844–2846

Published Sources

Agassiz, A., "A Visit to the Great Barrier Reef of Australia in the Steamer 'Croydon,' during April and May, 1896," *Bulletin of the Museum of Comparative Zoology at Harvard College,* vol. 28, no 4, (April 1898)

Agassiz, G. R., ed., *Letters and Recollections of Alexander Agassiz: With a Sketch of His Life and Work,* Boston and New York, Hougton Mifflin, 1913

Agnew, V. and J. Lamb, eds., *Criticism: A Quarterly for Literature and the Arts,* Special Issue: Extreme and Sentimental History, vol. 46, no. 3, Summer, 2004

Alexander, M., *Mrs. Fraser on the Fatal Shore,* London, Sphere Books, 1976

Allen, J. and P. Corris, eds., *The Journal of John Sweatman: A Nineteenth Century Surveying Voyage in North Australia and Torres Strait,* St. Lucia, University of Queensland Press, 1977

Altick, R. D., *Victorian Studies in Scarlet,* New York and London, Norton, 1970

Anderson, S., *Pelletier: The Forgotten Castaway of Cape York,* Melbourne, Melbourne Books, 2009, and revised edition, 2013

———. "Three Living Australians and the Société d'Anthropologie de Paris, 1885," in *Foreign Bodies: Oceania and the Science of Race 1750–1840,* eds., Bronwyn Douglas and Chris Ballard, Canberra, ANU Press, 2008

Arnold, M. L. and Nicole D. Fogarty, "Reticulate Evolution and Marine Organisms: The Final Frontier?," *International Journal of Molecular Sciences,* vol. 10, no. 9

Bach, J. P. S., *The Pearling Industry of Australia*, Canberra, Commonwealth of Australia, 1955

Baker, S., *The Ship: Retracing Cook's Endeavour Voyage*, London, BBC, 2002

Ballantyne, R. M., *The Coral Island*, London, Thames Publishing, n.d. [1858]

Banfield, E. J., *The Confessions of a Beachcomber* (1908), Rowville, Five Mile Press, 2006

———. "Dunk Island—Its General Characteristics," *Queensland Geographical Journal*, vol. 23, 1907–08

———. *Last Leaves from Dunk Island*, South Yarra, Angus and Roberson, 1925

———. *My Tropic Isle*, London and Leipsic, T. Fisher Unwin, 1912

———. *The Torres Strait Route: From Queensland to England*, Townsville, Thankful Willmett, 1885

———. *Tropic Days* (1917), e-books, University of Adelaide, 2010

Banks, J., *The Endeavour Journal of Joseph Banks*, 2 vols., edited by J. C. Beaglehole, Sydney, Trustees of Public Library of NSW in association with Angus and Robertson, 1963

Barker-Benfield, G. J., "Sensibility," in Iain McCalman, ed. *An Oxford Companion to the Romantic Age: British Culture, 1776–1832*, Oxford, Oxford University Press, 1999

Barrett, C., "The Great Barrier Reef and Its Isles: The Wonder and Mystery of Australia's World-Famous Geographical Feature," *National Geographic Magazine*, vol. 58 (1930)

———. *Koonwarra: A Naturalist's Adventures in Australia*, London, Oxford University Press, H. Milford, 1939

———. "On a Coral Isle: Barrier Reef Fish and Snakes," *Sydney Morning Herald*, August 8, 1928

———. "On a Coral Isle: Barrier Reef Theories," *Sydney Morning Herald*, August 8, 1928

———. "On a Coral Isle: Peering at Barrier Reef Secrets," *Sydney Morning Herald*, August 2, 1928

———. "On a Coral Isle: Scientific Research," *Sydney Morning Herald*, August 8, 1928

———. "Science in Shirt Sleeves," *Adelaide Register*, August 14, 1928

———. "Scientists in Wonderland," *Adelaide Register*, August 1, 1928

———. *The Sunlit Land: Wanderings in Queensland*, Melbourne, Cassell and Company, 1947

Bassett, M., *Behind the Picture: HMS Rattlesnake's Australia-New Guinea Cruise*, Melbourne, Oxford University Press, 1966

Beckett, J., *Torres Strait Islanders: Custom and Colonialism*, Cambridge, Cambridge University Press, 1989

Behrendt, S. D. et al., "Designing a Multi-Source Relational Database: 'Liverpool as Trading Port, 1700–1850,'" *International Journal of Maritime History*, vol. 24, no. 1 (June 2012), pp. 265–300

Bennett, I., *The Fringe of the Sea*, Sydney, Rigby Ltd, 1966

———. *The Great Barrier Reef*, Melbourne, Lansdowne, 1971

Bolton, G. C., *A Thousand Miles Away: A History of North Queensland to 1920*, Canberra, Jacaranda Press, 1970

Borschmann, G., "John Geoffrey Tracey, The Field Botanist," in *The People's Forest: A Living History of the Australian Bush*, Blackheath, NSW, People's Forest Press, 1999

Bowen, J. and M., *The Great Barrier Reef: History, Science, Heritage*, New York, Cambridge University Press, 2002

Bowler, P. J., *The Environmental Sciences*, London, Fontana, 1992

Brady, V., *South of My Days: A Biography of Judith Wright*, Sydney, Angus and Robertson, 1998

Brayshaw, H., *Well Beaten Paths: Aborigines of the Herbert-Burdekin District of North Queensland: An Ethnographic and Archaeological Study*, Townsville, James Cook University, 1990

Breslin, B., "Exterminate with Pride," *Aboriginal–European Relations in the Townsville-Bowen Region*, 1843–1869, Townsville, James Cook University, 1992

Brockett, W. E., *Narrative of a Voyage from Sydney to Torres Straits in Search of the Survivors of the Charles Eaton*, Sydney, Henry Bull, 1836

Brockway, L. H., *Science and Colonial Expansion: The Role of the British Royal Botanic Gardens*, New Haven, CT, and London, Yale University Press, 2002

Broome, R., *Aboriginal Australians: Black Responses to White Dominance 1788–2001*, Sydney, Allen & Unwin, 2001

Brown, B. E., "The Legacy of Professor John Stanley Gardiner, FRS, to Reef Science," *Notes and Records of the Royal Society*, vol. 61 (May 22, 2007)

Brown, R., *Nature's Investigator: The Diary of Robert Brown in Australia, 1801–1805*, compiled by T. G. Vallance, D. T. Moore, and E. W. Groves, Canberra, Australian Biological Resources Study, 2001

Browne, C. A., ed., *Letters and Extracts from the Addresses and Occasional Writings of J. Beete Jukes*, London, Chapman and Hall, 1871

Browne, J., *Charles Darwin: Voyaging*, London, Jonathan Cape, 1995

Burke, E., *A Philosophical Enquiry into the Origin of our Ideas of the Sublime and Beautiful*, London, R. and J. Dodsley, 1757

Carson, R., *The Sea*, London, MacGibbon and Kee, 1964

———. *Silent Spring*, Boston and New York, Houghton Mifflin, 1962

Chase, A., "Pama Malngkana," in Stephanie Anderson, *Pelletier: The Forgotten Castaway of Cape York*, Melbourne, Melbourne Books, 2009

Chisholm, A. H., *Birds and Green Places: A Book of Australian Nature Gossip*, London, J. M. Dent and Sons, 1929

Clare, P., *The Struggle for the Great Barrier Reef*, London, Collins, 1971

Clarke, P., and M. McKinney, *With Love and Fury: Selected Letters of Judith Wright*, Canberra, National Library of Australia, 2006

Colley, L., *Captives*, New York, Pantheon, 2002

Collins, P., "The Molecatcher's Daughter," *The Independent*, November 26, 2006

Cook, J., *An Account of a Voyage Round the World with a Full Account of the Voyage of the Endeavour in the Year MDCCLXX Along the East Coast of Australia*, ed. J. Hawkesworth, Brisbane, Smith and Paterson, 1969

———. *The Journals of Captain James Cook on His Voyages of Discovery*, vol. 1, ed. J. C. Beaglehole, Cambridge, published for the Hakluyt Society at University Press, 1955

Curtis, J., *SHIPWRECK of the STIRLING CASTLE, . . . the Dreadful Sufferings of the Crew, . . . THE CRUEL MURDER OF CAPTAIN FRASER BY THE SAVAGES [and] . . . the Horrible Barbarity of the Cannibals Inflicted upon THE CAPTAIN'S WIDOW, Whose Unparalleled Sufferings Are Stated by Herself, and Corroborated by the Other Survivors*, London, George Virtue, 1838

Dakin, W. J., *The Great Barrier Reef*, Sydney, Walkabout Pockets, 1968

Darian-Smith, K., R. Poignant, and K. Schaffer, *Captive Lives: Australian Captivity Narratives*, London, Chameleon Press, 1993

Darwin, C., *The Autobiography of Charles Darwin*, 1809–82, ed. N. Barlow, New York and London, Norton, 1958

Darwin, F., ed., *Life and Letters of Charles Darwin*, 3 vols., London, Murray, 1887

Davenport, C. B., "Alfred Goldsborough Mayor," *Memoirs of the National Academy of Science*, vol. 21 (1926)

Davidson, J., "Beyond the Fatal Shore: The Mythologization of Mrs. Fraser," *Meanjin*, vol. 3 (1990)

Desmond, A., *Huxley: From Devil's Disciple to Evolution's High Priest*, Reading, MA, Addison-Wesley, 1997

Dickens, C., *The Mystery of Edwin Drood* (1870), London, Folio Society, 1982

Dobbs, D., *Reef Madness: Charles Darwin, Alex Agassiz and the Meaning of Coral*, New York, Pantheon Books 2005

Douglas, B., *Across the Great Divide: Journeys in History and Anthropology*, Amsterdam, Harwood Academic Publishers, 1998

Embury, E. M., *The Great Barrier Reef*, Sydney, Shakespeare Head Press, 1933

Estensen, M., *The Life of Matthew Flinders*, Sydney, Allen & Unwin, 2002

Evans, R. and J. Walker, " 'These Strangers, Where Are They Going?' Aboriginal-European Relations in the Fraser Island and Wide Bay Region, 1770–1905," *Occasional Papers in Anthropology*, no. 8, March 1977

Farnfield, J., *Frontiersman: A Biography of George Elphinstone Dalrymple*, Melbourne, Oxford University Press, 1968

Fitzpatrick, M., "Enlightenment," in Iain McCalman, ed., *An Oxford Companion to the Romantic Age: British Culture, 1776–1832,* Oxford, Oxford University Press, 1999

Fleming, F., *Barrow's Boys,* London, Granta, 1998

Flinders, M., *Matthew Flinders: Personal Letters from an Extraordinary Life*, ed. by Paul Brunton, Sydney, Hordern House in association with the State Library of New South Wales, 2002

———. *A Voyage to Terra Australis . . . 1801, 1802 and 1803 in His Majesty's Ship The Investigator . . .* 2 vols., plus atlas, London, G. and W. Nicol, 1814

Forester, C. S., *Captain Hornblower, RN*, London, Michael Joseph, 1939

Frankel, E., *Bibliography of the Great Barrier Reef Province,* Canberra, Great Barrier Reef Marine Park Authority, 1978

Frost, A., *East Coast Country: A North Queensland Dreaming*, Melbourne, Melbourne University Press, 1996

Gammage, B., *The Biggest Estate on Earth: How Aborigines Made Australia*, Sydney, Allen & Unwin, 2011

Ganter, R., *The Pearl Shellers of Torres Strait: Resource Use and Decline, 1860s–1960s*, Melbourne, Melbourne University Press, 1994

Gascoigne, J., *Captain Cook: Voyager Between Worlds*, London, Hambledon Continuum, 2007

Gillett, K. and F. McNeill, *The Great Barrier Reef and Adjacent Isles*, Sydney, The Coral Press, 1959

Good, P., *The Journal of Peter Good, Gardener on Matthew Flinders Voyage to Terra Australis 1801–1803*, ed. Phyllis I. Edwards, London, British Museum, 1981

Goodman, J., *The Rattlesnake: A Voyage of Discovery to the Coral Sea*, London, Faber, 2005

Grant, J., *The Great Metropolis*, 2 vols., London, Saunders and Otley, 1837

Hamilton-Paterson, J., *Seven-Tenths: The Sea and Its Thresholds*, New York, Europa Editions, 2009

Harrison, A. J., *Savant of the Australian Seas: William Saville-Kent, 1842–1908*, Hobart, Tasmanian Historical Research Association, 2nd edition, n.d. [1997]

Hill, D., "The Great Barrier Reef Committee, 1922–82: The First Thirty Years," *Historical Records of Australian Science*, vol. 6, no. 1

Hingston Quiggin, A., *Haddon: The Head Hunter*, Cambridge, Cambridge University Press, 1942

Hoegh-Guldberg, O., *Climate Change, Coral Bleaching, and the Future of the World's Coral Reefs*, Greenpeace Report, 1999

Holmes, R., *Coleridge: Early Visions*, Harmondsworth, Penguin, 1990

Holthouse, H., *Ships in the Coral: Explorers, Wrecks and Traders of the Northern Australian Coast*, Sydney, Angus and Robertson, 1976

Horwitz, T., *Into the Blue: Boldly Going Where Captain Cook Has Gone Before*, Sydney, Allen & Unwin, 2002

Howard, R., *The Fabulist: The Incredible Story of Louis de Rougemont*, Sydney, Random House, 2006

Huxley, T. H., *Diary of the Voyage of the Rattlesnake*, ed. Julian Huxley, New York, Doubleday, 1936

————. "An Episcopal Trilogy" (1887), *Essays: Science and Christian Tradition*, New York, Appleton, 1896

————. "Scientific and Pseudo-Scientific Realism" (1887), *Essays: Science and Christian Tradition*, New York, Appleton, 1896

Ingleton, G. C., *Charting a Continent*, Sydney, Angus and Robertson, 1944

Ireland, J. *The Shipwrecked Orphans: A True Narrative of Four Years' Sufferings*, ed. Thomas Teller, New Haven, CT, Babcock, 1845

James, L., *Print and the People*, 1819–51, London, Viking, 1976

Jones, S., *Coral: A Pessimist in Paradise*, London, Little, Brown, 2007

Jukes, J. B., *Narrative of the Surveying Voyage of H.M.S. Fly . . . in the Torres Strait, New Guinea, and Other Islands of the Eastern Archipelago, during the years 1842–1846*, 2 vols., London, T. and W. Boone, 1847

Kyle, N., *A Greater Guilt: Constance Emilie Kent and the Road Murder*, Brisbane, Boolarong Press, 2009

Lavery, H. J., *Exploration North: A Natural History of Queensland*, Melbourne, Lloyd O'Neil, 1983

Lawrence, D., R. Kenchington, and S. Woodley, *The Great Barrier Reef: Finding the Right Balance*, Melbourne, Melbourne University Press, 2002

Lewis, C. M., *A Voyage to the Torres Strait in Search of the Survivors of the Charles Eaton, Which Was Wrecked upon the Barrier Reefs, in HM Colonial Schooner Isabella, C.M. Lewis Commander*, arranged by P. P. King, Sydney, E. H. Statham, 1837

Lewis, D., *A Wild History: Life and Death on the Victoria River Frontier*, Melbourne, Monash University Publishing, 2012

Lines, W. J., *Patriots: Defending Australia's Natural Heritage*, St. Lucia, University of Queensland Press, 2006

Lloyd, C., *Mr. Barrow of the Admiralty: A Life of Sir John Barrow*, London, Collins, 1970

Lock, A. C. C., *Destination Barrier Reef*, Melbourne, Georgian House, 1955

Loos, N., *Invasion and Resistance: Aboriginal-European Relations on the North Queensland Frontier, 1861–1897*, Canberra, Australian National University, 1982

Love, R., *Reefscape: Reflections on the Great Barrier Reef*, Washington, DC, Joseph Henry Press, 2001

Love, R., and N. Grasset, "Women Scientists and the Great Barrier Reef: A Historical Perspective," *Search*, vol. 15 (October–November 1984)

Macfarlane, R., *Mountains of the Mind: A History of a Fascination*, London, Granta, 2008

MacGillivray, J., *Narrative of the Voyage of HMS Rattlesnake . . . During the Years, 1846–1850*, 2 vols., London, Boone, 1852

Mack, J. D., *Matthew Flinders, 1774–1814*, Edinburgh and London, Thomas Nelson, 1966

Malouf, D., *Remembering Babylon*, Sydney, Vintage, 2009

Manion, J., *Paper Power in North Queensland: A History of Journalism in Townsville and Charters Towers*, Townsville, North Queensland Newspaper Company Ltd, 1982

Mayer, A. G., see A. G. Mayor

Mayor, A. G., "Alexander Agassiz, 1835–1910," *Popular Science Monthly*, vol. 77 (November 1910)

————. "Ecology of the Murray Island Coral Reef," *Papers from the Department of Marine Biology of the Carnegie Institution of Washington*, vol. 9, no. 213 (1918), Washington, DC

————. "An Expedition to the Coral Reefs of the Torres Straits," *Popular Science Monthly*, vol. 85 (September 1914)

McCalman, I., *Darwin's Armada: How Four Voyagers to Australasia Won the Battle for Evolution and Changed the World*, Melbourne, Penguin, 2010

————. *Radical Underworld: Prophets, Revolutionaries and Pornographers in London, 1795–1840*, Cambridge, Cambridge University Press, 1988

McKenna, M., *Looking for Blackfella's Point: An Australian History of Place*, Sydney, University of NSW Press, 2002

McLynn, F., *Captain Cook: Master of the Seas*, New Haven, CT, and London, Yale University Press, 2011

Melville, H. S., *The Adventures of a Griffin on a Voyage of Discovery. Written by Himself,* London, Bell and Daldy, 1867

Mendenhall, T. C., "Alfred Goldsborough Mayor," *Science*, new series, vol. 56, no. 1442 (August 18, 1922)

Merland, C., "Dix-sept ans chez les sauvages: Les aventures de Narcisse Pelletier" (1876), in Stephanie Anderson, *Pelletier: The Forgotten Castaway of Cape York*, Melbourne, Melbourne Books, 2009

Monkman, N., *From Queensland to the Barrier Reef,* New York, Doubleday and Co., 1958

——. *Quest of the Curly-Tailed Horses*, Sydney, Angus and Robertson, 1962

Moore, D. R., *Islanders and Aborigines at Cape York: An Ethnographic Reconstruction Based on the 1848–1850 "Rattlesnake" Journals of O. W. Brierly and Information Obtained from Barbara Thompson*, Canberra, Australian Institute of Aboriginal Studies, 1979

Morrill, J., *Seventeen Years Wandering Among the Aboriginals*, reprint of 1864 edn., Aboriginal Culture Series No. I, Virginia, NT, D. Welch, 2006

——. *Sketch of a Residence Among the Aboriginals of North Queensland*, Brisbane, Courier Office, 1863

Morton, B., "Charles Maurice Yonge," *Biographical Memoirs of the Royal Society*, vol. 38 (1992)

Morton-Evans, C. and M., *The Remarkable Life of Ellis Rowan*, Sydney, Simon and Schuster, 2008

Mulligan, M. and S. Hill, *Ecological Pioneers: A Social History of Australian Ecological Thought and Action*, Cambridge, Cambridge University Press, 2001

Murray, J., "Alexander Agassiz: His Life and Scientific Work," *Bulletin of the Museum of Comparative Zoology at Harvard College*, vol. 54, no. 3 (March, 1911)

Napier, S. E., *On the Barrier Reef* [1928], Sydney and London, Angus and Robertson, 1939 reprint

Nessy, A., "The Sea Has Many Voices: Profile of an Australian Woman Scientist," *Journal of Australian Studies*, vol. 38 (1992), pp. 41–50

Noonan, M., *A Different Drummer: The Story of E. J. Banfield, Beachcomber of Dunk Island*, St. Lucia, University of Queensland Press, 1986

Peake, J. H., *A History of the Burdekin*, Ayr, Shire of Ayr Council, 1951

Pollak, M. and M. MacNabb, *Hearts and Minds: Creative Australians and the Environment*, Sydney, Hale and Iremonger, 2000

Pollock, E. F., *Expeditions to Great Barrier Reef Islands Off the Queensland Coast*, Strathfield, NSW, E.F. Pollock, n.d.

Powell, A., *Northern Voyagers: Australia's Monsoon Coast in Maritime History*, Melbourne, Australian Scholarly Publishing, 2010

Rae-Ellis, V., *Louisa Anne Meredith: A Tigress in Exile*, Hobart, Blubber Head Press, 1979

Reid, J. C., *Bucks and Bruisers, Pierce Egan and Regency England*, London, Routledge, 1971

Reynolds, H., "Aboriginal Resistance in Queensland," *Australian Journal of Politics and History*, vol. 22 (April, 1976)

——. *North of Capricorn: The Untold Story of Australia's North*, Sydney, Allen & Unwin, 2003

——. *The Other Side of the Frontier*, Townsville, James Cook University, 1981

Richards, H. C., "Problems of the Great Barrier Reef," *Queensland Geographical Journal*, vol. 37 (1922)

Roland, B., *The Eye of the Beholder*, Sydney, Hale and Iremonger, 1984

Rose, D. B., "Writing Place," in Ann Curthoys and Ann McGrath, eds., *Writing Histories: Imagination and Narration*, Monash e-books, 2009, chapter 8.

Rougemont, L. de, *The Adventures of Louis de Rougemont*, Project Gutenberg e-book, April 17, 2007, transcribed from George Newnes edition, London, 1899

Roughley, T. C., *Wonders of the Great Barrier Reef*, Sydney and London, Angus and Robertson, 1945

Russell, F., "Sheina Macalister Marshall, 20 April 1896–7 April 1977," *Biographical Memoirs of Fellows of the Royal Society*, vol. 24 (1978)

Saville-Kent, W., *The Great Barrier Reef of Australia: Its Products and Potentialities* (1893), facsimile edition, Melbourne, Currey, O'Neill, 1972

———. *A Manual of the Infusoria: Including a Description of All Known Flagellate, Ciliate and Tentaculiferous Protozoa*, 3 vols., London, D. Bogue, 1880–81

———. *The Naturalist in Australia*, London, Chapman and Hall, 1897

———. "Preliminary Observations on a Natural History Collection Made on the Surveying Cruise of HMS Myrmidon," *Papers and Proceedings of the Royal Society of Queensland*, vol. 6 (1889)

Schaffer, K., "Captivity Narratives and the Idea of Nation," in *Captured Lives: Australian Captivity Narratives*, eds., Kate Darian Smith, Roslyn Poignant and Kay Schaffer, Robert Menzies Centre for Australian Studies, London, 1993

———. *In the Wake of First Contact: The Eliza Fraser Stories*, Cambridge, Cambridge University Press, 1995

Simmons, J. C., *Castaway in Paradise: The Incredible Adventures of True-Life Robinson Crusoes*, New York, Sheridan House, 1993

Single, J., *The Torres Strait: People and History*, St. Lucia, University of Queensland Press, 1989

Smith, K. V., *Mari Nawi: Aboriginal Odysseys*, New South Wales, Rosenberg, 2010

Smyth, D., *Aboriginal Maritime Culture in the Far Northern Section of the Great Barrier Reef Marine Park*, Townsville, Great Barrier Reef Marine Park Authority, 1995

———. *Understanding Country: The Importance of Land and Sea in Aboriginal and Torres Strait Islander Societies*, Canberra, Australian Government Publishing Service, 1994

Stephens, L. D. and D. R. Calder, *Seafaring Scientist: Alfred Goldsborough Mayor, Pioneer in Marine Biology*, Columbia, University of South Carolina Press, 2006

Summerscale, K., *The Suspicions of Mr Whicher, or The Murder at Road Hill House*, London, Bloomsbury, 2009

Taylor, B., *Cruelly Murdered: Constance Kent and the Killing at Road Hill House*, London, Souvenir, 1979

Teichgraeber, R. F., *Sublime Thoughts/Penny Wisdom: Situating Emerson and Thoreau in the American Market*, Baltimore, Johns Hopkins, 1995

Thomas, N., *Cook: The Extraordinary Voyages of Captain James Cook*, New York, Walker and Co., 2003

———. "Exploration," in Iain McCalman, ed., *An Oxford Companion to the Romantic Age: British Culture, 1776–1832*, Oxford, Oxford University Press, 1999

Thoreau, H. D., *Walden; or, Life in the Woods* (1854), New York, Dover, 1995

Todd, J., *Sensibility: An Introduction*, London and New York, Methuen, 1986

Tracey, G., "Len Webb: Pioneer in Ecology of the Rainforests of Australia," *Australian Science Magazine*, vol. 4 (1988)

Vallance, T. G., D. T. Moore, and E. W. Groves, *Nature's Investigator: The Diary of Robert Brown in Australia, 1801–1805*, Canberra, Australian Biological Reseources Study, 2001

Veron, J. E. N., "1996 AMSA Silver Jubilee Awardee, Dr. Charlie Veron," reprinted from *Golden Anniversary Issue of the Atoll Research Bulletin*, vol. 494 (2001)

———. *Corals in Space and Time: The Biogeography and Evolution of the Scleractinia*, Sydney, University of NSW Press, 1995

———. "Is the Great Barrier Reef on Death Row?," Lecture at the Royal Society, London, July 6, 2009

———. *A Reef in Time: The Great Barrier Reef From Beginning to End*, Cambridge, MA, Harvard University Press, 2008

———. "The Songlines Conversations: John (Charlie) Veron," interviewed by Gregg Borschmann, *Big Ideas*, ABC Radio, August 6, 2006

Veth, P., P. Sutton, and M. Neale, *Strangers on the Shore: Early Coastal Contacts in Australia*, Canberra, National Museum of Australia Press, 2008

Wain, B., "The Bingal Bay Bastard," *Nation* (May 1, 1971)

Walker, J. B., *Prelude to Federation*, Hobart, Blubber Head Press, 1976

Warren, R. J., *Wildflower: The Barbara Crawford Thompson Story*, Brisbane, R. J. Warren, revised 3rd edition, 2012

Webb, L., "Notes for an Address to the Students at Kendron Teachers College," *Wildlife*, no. 7, June 1967

Webb, L. and J. W. T. Williams, "Synecology—Cinderella Finds Her Coach," *New Scientist*, no. 59, July 26, 1973

Wemyss, T., *Narrative of the Melancholy Shipwreck of the Charles Eaton*, Stockton, W. Robinson, 1837

Werkmeister, L. T., *The London Daily Press, 1772–1792*, Lincoln, University of Nebraska Press, 1963

———. *A Newspaper History of England, 1792–93*, Lincoln, University of Nebraska Press, 1967

White, P., *A Fringe of Leaves*, London, Vintage, 1997

Wildlife Preservation Society of Queensland Historical Papers, "Catalyst for Action: Formation of a Conservation Society," *WPSQ Historical Papers: Monograph 1*, July 2006

———. "Heart and Mind: WPSQ Finding Directions in the 60s," *WPSQ Historical Papers: Monograph 2*, May 2008

Wilson, K., *Island Race: Englishness, Empire and Gender in the 18th Century*, London, Routledge, 2003

Wilson, T. B., *Narrative of a Voyage Round the World: Comprehending an Account of the Wreck of the Ship, Governor Ready in the Torres Strait*, London, Gilbert and Piper, 1835

Woodford, J., *The Great Barrier Reef: In Search of the Real Reef*, Sydney, Macmillan, 2010

Woodward, R. S., "Alfred Goldsborough Mayor," *Science*, vol. 56, no. 1438 (July 21, 1922)

Worrell, E., *The Great Barrier Reef*, Sydney, Angus and Robertson, 1966

Wright, J., *Because I Was Invited*, Melbourne, Oxford University Press, 1975

———. *Collected Poems, 1942–85*, Sydney, Angus and Robertson, 1994

———. *The Coral Battleground*, Sydney, Angus and Robertson, 1996

———. "The Quiet Crisis of Our Time," *Outlook*, no. 3, June 1969

———. "The Role of Public Opinion in Conservation," in *The Last of Lands: Conservation in Australia*, L. J. Webb, D. Whitelock, J. Le Gay Brereton, eds., Milton, Queensland, Jacaranda, 1969

Yonge, C. M., *A Year on the Great Barrier Reef: The Story of the Corals and the Greatest of Their Creations*, London and New York, Putnam, 1930

ACKNOWLEDGMENTS

I first fell in love with the Great Barrier Reef on a hapless BBC reenactment that I describe briefly in the Prologue. This also proved to be one of the more traumatic times of my life, and I would like to thank my friends and colleagues Jonathan Lamb, Vanessa Agnew, and Alex Cook for making it bearable, and so contributing over the long haul to the genesis of this book.

Shaping, researching, and writing *The Reef* has proved to be another type of long haul, one in which I was always sustained by the keen perceptions, good humor, and pithy editing skills of my loving wife, Kate Fullagar. And, as with previous books, my hardworking agent Mary Cunnane has offered unstinting wisdom, support, and guidance, ably assisted by Kathleen Anderson in New York and Natasha Fairweather in London.

My warmest thanks and admiration also go to senior science editor Amanda Moon and the wonderful team at Farrar, Straus and Giroux, including Daniel Gerstle and Laird Gallagher. It is a pure pleasure to work with such clever, professional, and caring people. The marketing and advertising of books and media has become an increasingly complex and demanding process these days, and for expertly managing this side of things I also thank Jeff Seroy.

I owe so much to my longtime research assistant and adviser, Katherine Anderson, who has managed with her usual calm efficiency to give birth both to this book and to a gorgeous baby girl, Corisande.

The writing and research of *The Reef* has been pursued in tandem with filming and other forms of digital production, with the aim that these media should feed off and fertilize one another. This fusion has been made possible by the friendship and brilliance of Mike Bluett of Northern Dogs Television and Digital. Working with him and learning from him has been one of the most energizing experiences of my late academic career. Among other things he has convinced me that digital and written histories can, and must, learn to work together in a creative symbiosis.

Around us we also gathered a splendid team of digital experts and friends: Sam Wilson, film editor; James Stewart, sound engineer; Dean Miller, cameraman; and Andrew

McCalman on video and stills, as well as Keren Moran and Noa Peer, Web creators extraordinaire of the Sydney digital company Spring in Alaska.

Our visits to Reef sites in order to film introduced us to a range of special individuals for whom working to protect the Reef and its peoples remains urgent, unfinished business. Here I would like particularly to mention Alberta Hornsby, a Guugu Yimithirr knowledge custodian of Hopevale and Cooktown, and a historian of great passion, sagacity, and balance. I am deeply in her debt. At Dunk Island we were inspired by the local knowledge of a longtime island inhabitant and nature lover, Susi Kirk, an inheritor of the mantle of Ted Banfield if ever there was one. At Lockhart River we were treated with extraordinary generosity: Paul Piva loaned us one of his four-wheel-drive vehicles and refused to take payment, while Wayne Butcher, the energetic mayor and community leader, provided us with a boat and his scarce gas supply, and then gave his Uutaalnganu-descended staff time off from work to talk with us about their ancestor Anco and the days when their community still lived in their Sandbeach country near Night Island.

An early visit to Rivendell, the home of coral scientists John "Charlie" Veron and Mary Stafford-Smith, introduced me to Charlie's incredible life and work; and he later also entrusted me with a moving private memoir written for his children. This began a relationship that has given vital shape and purpose to my book, though he is not to blame for my mistakes. The Great Barrier Reef and our planet owe these two public-spirited scientists an incalculable debt.

If one is lucky universities can provide supportive and inspirational environments of a different kind. Friendships and informal conversations with colleagues at Sydney and other universities have been more sustaining than these scholars can know. Here I would particularly like to thank Shane White, Mike McDonnell, David Schlosberg, Alison Bashford, Mark McKenna, Jodi Frawley, Julia Horne, Duncan Ivison, Kirsten McKenzie, Jude Philp, Leah Lui-Chivezhe, Michael Davis, Clare Corbould, Ann Curthoys, John Docker, Barbara Caine, Leigh Boucher, Nicholas Thomas, Jim Chandler, John Barrell, Harriet Guest, Jon Mee, Libby Robin, and my special fount of science knowledge, Lachlan McCalman.

The actual funding to enable research for this book and its associated digital productions came through the generosity of the Australian Research Council, whose Linkage grant also brought me into collaboration with Michael Westaway of the Queensland Museum and with my old friends Michael Crayford and Nigel Erskine of the Australian National Maritime Museum. Crucially it also cemented a collaborative partnership with John Mullen, director of Silentworld Foundation, and his wife, Jackie, which has ripened into a warm friendship. John's knowledge of and passion for early Australian maritime history, as well as for archaeological diving in sometimes perilous circumstances, and for collecting, preserving, and displaying vital objects of Australian heritage stands as a salutary example to historical and museum professionals everywhere.

As usual I have depended on the generosity and expertise of librarians and archivists for my intellectual infrastructure of manuscripts and illustrations. I would like to thank the Fisher Library, University of Sydney; the National Library of Australia; the Mitchell Library, State Library of New South Wales; the Australian Museum, Sydney; the State Russian Museum, St. Petersburg; the Royal Historical Society of Queensland; the John Oxley Library, State Library of Queensland; the Fryer Library, University of Queensland; the James Cook University Library; the Griffith University Library; the Australian National Maritime Museum; the Archives of the Ernst Mayr Library of the Museum of Comparative Zoology, Harvard University; the Australian Marine Conservation Society; the Wildlife Preservation Society of Queensland; and the Library of the Australian Institute of Marine Science, Townsville.

Finally I would like to acknowledge my intellectual debts to those historians on whose work I have depended. We all stand on the shoulders of our predecessors and none more so than me. The late James and Margarita Bowen are the great pioneers of Barrier Reef history, and I have followed in their wake. Bill Gammage's extraordinary new book *The Biggest Estate on Earth* taught me to understand Indigenous ecologies in wholly new ways. Reading Stephanie Anderson's wonderful translation and study of the life of Narcisse Pelletier—enriched by the anthropological expertise of Athol Chase—was how I came to know about Anco, the most fascinating Reef castaway of all. I thank Stephanie for her generosity in sharing her knowledge and insight with me.

Likewise I could not have written about early Australian marine and maritime history without the writings of Alan Frost and John Gascoigne; about James Cook and Joseph Banks without the great texts of J. C. Beaglehole; about Eliza Fraser without Kay Schaffer and Jim Davidson; about Barbara Thompson without Ray Warren's research and D. R. Moore's marvelous transcripts and anthropological analyses; about William Saville-Kent without the research of A. J. Harrison, Kate Summerscale, and Noreen Kyle; and about Alex Agassiz and Alfred Mayor without the prior works of D. Dobbs, L. D. Stephens, and D. R. Calder. Phoebe Ford located, copied, and provided me with valuable documents from the Busst papers; and Emma Dortins allowed me to quote from her recent PhD thesis, which contains important original information on the castaway James Morrill.

My grasp of the science and ecology of coral reefs and corals—such as it is—would have been far worse without Charlie Veron's brilliant *A Reef in Time*, and the sparkling wit and erudition of the University College London geneticist and science writer Steve Jones. And nobody can write about northern Australian Aboriginal history without immersing themselves in the writings of Henry Reynolds, or hope to understand the early history and culture of the Torres Strait without the work of Jeremy Beckett. My thanks to you all.

INDEX

A Note About the Author

Iain McCalman is a fellow of the Royal Historical Society, a historian, a social scientist, and an explorer. He is the author of *Darwin's Armada*, *The Last Alchemist*, and *Radical Underworld*. He is a professor of history at the University of Sydney.